NUMBER SENSE

NUMBER SENSE

HOW TO USE BIG DATA TO YOUR ADVANTAGE

KAISER FUNG

New York   Chicago   San Francisco   Athens   London   Madrid   Mexico City
Milan   New Delhi   Singapore   Sydney   Toronto

1 2 3 4 5 6 7 8 9 0   QFR/QFR   1 9 8 7 6 5 4 3

ISBN: 978-0-07-179966-9
MHID: 0-07-179966-4

e-ISBN: 978-0-07-179967-6
e-MHID: 0-07-179967-2

McGraw-Hill books are available at special quantity discounts to use as premiums and sales promotions, or for use in corporate training programs. To contact a representative, please visit the Contact Us page at www.mhprofessional.com.

This book is printed on acid-free paper.

Contents

Acknowledgments

I owe a great debt to readers of *Numbers Rule Your World* and my two blogs, and followers on Twitter. Your support keeps me going. Your enthusiasm has carried over to the McGraw-Hill team, led by Knox Huston. Knox shepherded this project while meeting the demands of being a new father. Many thanks to the production crew for putting up with the tight schedule. Grace Freedson, my agent, saw the potential of the book.

Jay Hu, Augustine Fou, and Adam Murphy contributed materials that made their way into the text. They also reviewed early drafts. The following people assisted me by discussing ideas, making connections or reading parts of the manuscript: Larry Cahoon, Steven Paben, Darrell Phillipson, Maggie Jordan, Kate Johnson, Steven Tuntono, Amanda Lee, Barbara Schoetzau, Andrew Tilton, Chiang-ling Ng, Dr. Cesare Russo, Bill McBride, Annette Fung, Kelvin Neu, Andrew Lefevre, Patty Wu, Valerie Thomas, Hillary Wool, Tara Tarpey, Celine Fung, Cathie Mahoney, Sam Kumar, Hui Soo Chae, Mike Kruger, John Lien, Scott Turner, Micah Burch, and Andrew Gelman. Laurent Lheritier is a friend whom I inadvertently left out last time. The odds are good that the above list is not complete, so please accept my sincere apology for any omission.

Double thanks to all who took time out of their busy lives to comment on chapters. A special nod to my brother Pius for being a willing subject in my experiment to foist Chapter 8 on non-sports fans.

This book is dedicated to my grandmother, who sadly will not see it come to print. A brave woman who grew up in tumultuous times, she taught herself to read and cook. Her cooking honed my appreciation for food, and since the field of statistics borrows quite a few culinary words, her influence is felt within these pages.

New York, April 2013

List of Figures

Prologue

If you were responsible for marketing at America West Airlines, you faced a strong headwind as 1990 winded down. The airline industry was going into a tailspin, as business travel plummeted in response to Operation Desert Storm. Fuel prices spiked as the economy slipped into recession. The success of the recent past, your success growing the business, now felt like a heavy chain around your neck. Indeed, 1990 was a banner year for America West, the upstart airline founded by industry veteran Ed Beauvais in 1983. It reached a milestone of $1 billion in revenues. It also became the official airline of the Phoenix Suns basketball team. When the U.S. Department of Transportation recognized America West as a "major airline," Beauvais's Phoenix project had definitively arrived.

Rival airlines began to drop dead. Eastern, Midway, Pan Am, and TWA were all early victims. America West retrenched to serving only core West Coast routes; chopped fares in half, raising $125 million and holding a lease on life. But since everyone else was bleeding, the price war took no time to reach your home market of Phoenix. You were seeking a new angle to persuade travelers to choose America West when your analyst came up with some sharp analysis about on-time performance. Since 1987, airlines have been required by

FIGURE P-1 America West Had a Lower Flight Delay Rate, Aggregate of Five West Coast Airports

	Alaska	America West
Total Flights	3,775	7,225
Total Delays	501	787
Proportion Delayed	13%	11%

the Department of Transportation to submit flight delay data each month. America West was a top performer in the most recent report. Only 11 percent of your flights arrived behind schedule, compared to 13 percent of flights of Alaska Airlines, a competitor of comparable size which also flew mostly West Coast routes (see Figure P-1).

Possible story lines for new television ads like the following flashed in your head:

> Guy in an expensive suit walks out of a limousine, gets tagged with the America West sticker curbside, which then transports him as if on a magic broom to his destination, while wide-eyed passengers looked on with mouths agape as they argued with each other in the airport security line. Meanwhile, your guy is seen shaking hands with his client, holding a signed contract and a huge smile, pointing to the sticker on his chest.

As it turned out, there would be no time to do anything. By the summer of 1991, America West declared bankruptcy, from which it emerged three years later after restructuring.

But so be it, as you'd just dodged a bullet. If you had asked the analyst for a deeper analysis, you would have found an unwelcome surprise. Take a look at Figure P-2.

Did you see the problem? While the average performance of America West beat Alaska's, the finer data showed that Alaska had fewer delayed flights at each of the five West Coast

FIGURE P-2 Alaska Flights Had Lower Flight Delay Rates Than
America West Flights at All Five West Coast Airports

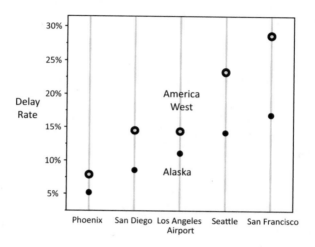

airports. Yes, look at the numbers again. The proportion of
delayed flights was higher than Alaska's at San Francisco, at
San Diego, at Los Angeles, at Seattle, and even at your home
base of Phoenix. Did your analyst mess up the arithmetic?
You checked the numbers, and they were correct.

I'll explain what's behind these numbers in a few pages.
For now, take my word that the data truly supported both of
these conclusions:

1. America West's on-time performance beat Alaska's on
   average;
2. The proportion of America West flights that were on time
   was lower than Alaska's at each airport.

(Dear Reader, if you're impatient, you can turn to the end of
the Prologue to verify the calculation.) Now, this situation is
unusual but not that unusual. One part of one data set does
sometimes suggest a story that's incompatible with another
part of the same data set.

I wouldn't blame you if you are ready to burn this book, and vow never to talk to the lying statisticians ever again. Before you take that step, realize that we live in the new world of Big Data, where there is no escape from people hustling numbers. With more data, the number of possible analyses explodes exponentially. More analyses produce more smoke. The need to keep our heads clear has never been more urgent.

Big Data: This is the buzzword in the high-tech world, circa early 2010s. This industry embraces two-word organizing concepts in the way Steven Seagal chooses titles for his films. Big Data is the heir to "broad-band" or "wire-less" or "social media" or "dot com." It stands for lots of data. That is all.

The McKinsey Global Institute—part of the legendary consulting firm McKinsey & Company—talks about "data sets whose size is beyond the ability of typical database software tools to capture, store, manage, and analyze." These researchers regarded "bigness" as a few dozen terabytes up to thousands of terabytes per enterprise, as of 2011 when they issued one of the first "Big Data" reports.

My idea of Big Data is more expansive than the industry standard. The reason why we should care is not more data, but more data *analyses*. We deploy more people producing more analyses more quickly. The true driver is not the amount of data but its availability. If we want to delve into unemployment or inflation or any other economic indicator, we can obtain extensive data sets from the Bureau of Labor Statistics website. If a New York resident is curious about the "B" health rating of a restaurant, he or she can review the list of past violations on the Department of Health and Mental Hygiene's online database. When the sudden acceleration crisis engulfed Toyota several years ago, we learned that the National Highway Traffic Safety Administration maintains an open repository of safety complaints by drivers. Since the early 1990s, anyone can download data on the performance of stocks, mutual funds, and other financial investments from a variety of websites such as Yahoo! Finance and E*Trade. Sometimes, even businesses get in on the act, making proprietary data public. In 2006, Netflix, the DVD-plus-streaming-media company, released 100 million movie ratings and enlisted scientists to improve its predictive algorithms. The availability of data has propelled the fantasy sports business to new heights, as players study statistics to gain an edge. The data which once appeared in printed volumes is now disseminated on the Internet in the form of spreadsheets. With so much free and easy data, there is bound to be more analyses.

Bill Gates is a classic American success story. A super-smart kid who dropped out of college, he started his own company, developed software that would eventually run 90 percent of the world's computers, made billions while doing it, and then retired and dedicated the bulk of his riches to charitable causes. The Bill & Melinda Gates Foundation is justly celebrated for bold investments in a number of areas, including malaria prevention in developing countries, high school reform in the United States, and HIV/AIDS research. The Gates Foundation has a reputation for relying on data to make informed decisions.

But this doesn't mean they don't make any mistakes. Gates threw his weight behind the small schools movement at the start of the millennium, pumping hundreds of millions of dollars into selected schools around the country. Exhibit A at the time was the statistical finding that *small schools accounted for a disproportionate share of the nation's top performing schools*. For example, 12 percent of the Top 50 schools in Pennsylvania ranked by fifth-grade reading scores were small schools, four times what would have been expected if achievement were unrelated to school size. Having identified size as the enemy—with 100 students per grade level as the tolerable limit—the Gates Foundation designed a reinvention plan around breaking up large schools into multiplexes.

For example, in the 2003 academic year, the 1,800 students of Mountlake Terrace High School in Washington found themselves assigned to one of five small schools, with names such as The Discovery School, The Innovation School, and The Renaissance School, all housed in the same building as before. Tom Vander Ark, the executive director of education at the Gates Foundation, explained his theory: "Most poor kids go to giant schools where nobody knows them, and they get shuffled into dead-end tracks. . . .Small schools simply produce an environment where it's easier to create a positive climate, high expectations, an improved curriculum, and better teaching [than large schools]."

Ten years later, the Gates Foundation made an about-turn. It no longer sees school size as the single solution to the student achievement problem. It's interested in designing innovative curriculums and promoting quality of teaching. Careful research studies, commissioned by the Gates Foundation, concluded that the average academic achievement of the reinvented schools was not better, and in some cases, was even worse.

Statistician Howard Wainer, who spent the better part of his career at Educational Testing Services, complained that the multimillion-dollar mistake was avoidable. In the same analysis of Pennsylvania schools referred to above, Wainer revealed that small schools accounted for 12 percent of the Top 50, and also 18 percent of the Bottom 50. So, small schools were overrepresented at both ends of the distribution. Depending on which part of the data is being highlighted, the analyst comes to contradictory conclusions. We saw a similar case in the study of flight delay. The key isn't how much data is analyzed, but how.

The Gates Foundation's story makes another point. Data analysis is tricky business, and neither technocrats nor experts have a monopoly on getting it right. No matter how brilliant someone is, there is always a margin of error, because no one has full information. "It's published in a top journal" is used as an excuse to mean "Don't ask questions." In the world of Big Data, only fools take that attitude. You have heard of many studies purported to link certain genes with certain diseases, from Parkinson's to hypertension. Are you aware that only 30 percent of these peer-reviewed and peer-approved findings of genetic associations could be confirmed by subsequent research? The rest are false-positive results. The reporters who have hyped the original findings almost never publish errata when they are overturned. That said, I expect experts, on average, to deliver a better quality of analysis.

If Wainer had done the original work on small schools, he would have taken a broad view of the data, and concluded that school size was a red herring. The evidence did not fit the theory, even if the theory that students benefit from individual attention has strong intuitive appeal. If the correlation between school size and achievement score were to exist, it would still have been insufficient to conclude that school size is *a cause*, or *the cause*, of the effect. (The challenge of causal data analysis is the topic of Chapter 2 of my previous book, *Numbers Rule Your World.*)

Big Data has essentially nothing to say about causation. It's a common misconception that an influx of data flushes cause—effect from its hiding place. Consider the clickstream, the click-by-click tracking of Web surfers frequently held up by digital marketers as causal evidence of their success. What stronger proof do you need than tying a final sale to a customer clicking on a banner ad or a search ad? The reality is far from tidy. Say, I clicked on a banner ad for the Samsung Galaxy but later left the phone in a shopping cart. Seven days later, I watched and loved their Apple-bashing commercial; I returned to the store and finalized the purchase. Not only would the analyst dissecting the Web logs miss the true cause of my action, but he would make a false-positive error by tying the purchase to the banner ad as that would be all he could see. This hiccup is uneventful in the life of a typical Web analyst. Here are some other worries:

- The number of verified transactions never equals the number of recorded clicks.
- Some transactions cannot be traced to any click, while others are claimed by multiple clicks.
- A slice of sales appeared to have arrived a few seconds *before* the attributed clicks.
- Some customers supposedly pressed on a link inside an e-mail without having opened it.
- The same person may have clicked one ad a hundred times within five minutes.

Web logs are a messy, messy world. If two vendors are deployed to analyze traffic on the same website, it is guaranteed that their statistics would not reconcile, and the gap can be as high as 20 or 30 percent.

Big Data means more analyses, and also more *bad* analyses. Even experts and technical gurus have their pants-are-unzipped moments. Some bad stuff is fueled by hurtful intentions of shady characters, but even well-meaning analysts can be tricked by the data. Consumers must be extra discerning in this data-rich world.

Data gives theory legitimacy. But every analysis also sits on top of theory.

Bad theory cannot be saved by data. Worse, bad theory and bad data analysis form a combustible mix. Republican pollsters who played with fire were scalded during the 2012 Presidential election, and it happened so swiftly that Karl Rove, the prominent political consultant, famously lost his head on live television when Fox News called Ohio, ergo the election for President Obama, at half-past eleven on the East Coast. Rove insisted that Ohio was not a done deal, forcing the host Megyn Kelly to corner the number crunchers in a back room for an "interrogation," in which she learned that they were "99.95 percent confident" about the disputed call.

Rove, as well as many prominent Republican pundits such as George Will, Newt Gingrich, Dick Morris, Rick Perry, and Michael Barone had predicted their candidate, Mitt Romney, would win the election handily. They had poll data to buttress their case. However, if you read *FiveThirtyEight*, the blog of Nate Silver, the *New York Times* guru of polls, you might have been wondering what the GOP honchos were smoking. For example, a selection of polls conducted in September 2012 indicated a comfortable lead of about 4 percentage points for President Obama (Figure P-3).

The immediate reaction from Romney's camp after his defeat was shock. They had projected a victory using apparently a different set of data, something that probably looked more like the data in Figure P-4 than the data in Figure P-3.

This second data set was the work of Dean Chambers, who runs a rival website to Nate Silver's called UnskewedPolls.com,

FIGURE P-3 National Polls on the 2012 U.S. Presidential Election: Includes Polls Conducted in September 2012
[*Source*: RealClearPolitics.com and UnskewedPolls.com]

POLL	DATES	OBAMA %	ROMNEY %	SPREAD
IBD/CSM/TIPP	9/4 - 9/9	46	44	Obama +2
CNN/Opinion Research	9/7 - 9/9	52	46	Obama +6
ABC News/Wash Post	9/7 - 9/9	49	48	Obama +1
Democracy Corps (D)	9/8 - 9/12	50	45	Obama +5
CBS News/NY Times	9/8 - 9/12	49	46	Obama +3
FOX News	9/9 - 9/11	48	43	Obama +5
NBC News/Wall St. Jrnl	9/12 - 9/16	50	45	Obama +5
Monmouth/SurveyUSA/Braun	9/13 - 9/16	48	45	Obama +3
Reason-Rupe/PSRAI	9/13 - 9/17	52	45	Obama +7
AVERAGE				*Obama +4*

FIGURE P-4 Re-weighted National Polls on the 2012 U.S.
Presidential Election: September 2012. [*Source*: UnskewedPolls.
com and RealClearPolitics.com]

POLL	DATES	OBAMA %	ROMNEY %	SPREAD	SPREAD
			Unskewed		*Unadjusted*
IBD/CSM/TIPP	9/4 - 9/9	41	50	Romney +9	Obama +2
CNN/Opinion Research	9/7 - 9/9	45	53	Romney +8	Obama +6
ABC News/Wash Post	9/7 - 9/9	45	52	Romney +7	Obama +1
Democracy Corps (D)	9/8 - 9/12	43	52	Romney +9	Obama +5
CBS News/NY Times	9/8 - 9/12	44	51	Romney +7	Obama +3
FOX News	9/9 - 9/11	45	48	Romney +3	Obama +5
NBC News/Wall St. Jrnl	9/12 - 9/16	44	51	Romney +7	Obama +5
Monmouth/SurveyUSA/Braun	9/13 - 9/16	45	50	Romney +5	Obama +3
Reason-Rupe/PSRAI	9/13 - 9/17	45	52	Romney +7	Obama +7
AVERAGE				*Romney +7*	*Obama +4*

which became a darling of the Republican punditry in the run-up to November 6. Chambers' numbers showed a sizable Romney lead in each poll, averaging 7 percentage points. What led him from minus 4 to plus 7 percentage points was a big serving of theory, and a pinch of bad data.

Chambers' *theory* was that there would be a surge in enthusiasm among Republican voters in the 2012 election, reflecting their unhappiness with the sluggish economic recovery and the disastrous jobs market (the topic of Chapter 6). Polling firms generally report results for *likely voters* only, which means the data incorporates a model of who is likely to vote. Chambers alleged that the likely-voter model was biased against Republicans as it did not account for the theorized jolt in red fever.

He set out to "unskew" the polling data. Needing a different way of estimating the party affiliation of likely voters, he turned to Rasmussen Reports, one of the less accurate polling firms in the business. Rasmussen polls collect party identification information via a prerecorded item on their auto dialer:

"If you are a Republican, press 1.

If a Democrat, press 2.

If you belong to some other political party, press 3.

If you are independent, press 4.

If you are not sure, press 5."

Here is where bad data entered the mix. Chambers reweighted results from other polls that he alleged undercounted likely Republican voters. By doing this, he also assumed that respondents to other polls mirrored the Rasmussen sample. After this adjustment, every poll foretold a Romney victory that never came to pass. Eventually, exit polls would estimate that 38 percent of voters were Democrats, 6 percentage points more than self-identified Republicans, annihilating Chambers' theory. Incidentally, polling firms do not have to guess who the likely voters are—they pose the question directly so that respondents "self-select" into the category.

In analyzing data, there is no way to avoid having theoretical assumptions. Any analysis is part data, and part theory. Richer data lends support to many more theories, some of which may contradict each other, as we noted before. But richer data does not save bad theory, or rescue bad analysis. The world has never run out of theoreticians; in the era of Big Data, the bar of evidence is reset lower, making it tougher to tell right from wrong.

People in industry who wax on about Big Data take it for granted that more data begets more good. Does one have to follow the other?

When more people are performing more analyses more quickly, there are more theories, more points of view, more complexity, more conflicts, and more confusion. There is less clarity, less consensus, and less confidence.

America West marketers could claim they had the superior on-time record relative to Alaska Airlines by citing the aggregate statistics of five airports. Alaska could counterclaim it had better timeliness by looking at airport-by-airport comparisons. When two conflicting results are on the table, no quick conclusion is possible without verifying the arithmetic, and arbitrating. The key insight in the flight delay data is the strong influence of the port of arrival, more so than the identity of the carrier. Specifically, flights into Phoenix have a much smaller chance of getting delayed than those into Seattle, primarily due to the contrast in weather. The home base of America West is Phoenix while Alaska has a hub in Seattle. Thus, the average delay rate for Alaska flights is heavily weighted toward a low-performing airport while the opposite is true for America West. The port-of-arrival factor hides the carrier factor. This explains the so-called *Simpson's Paradox* (Figure P-5).

FIGURE P-5 Explanation of Simpson's Paradox in Flight Delay Data

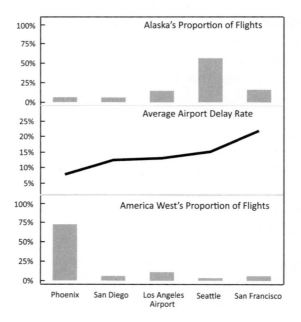

The airline analysis only uses the four entities: carrier, port of arrival, number of flights, and frequency of delays. Many more variables are available, such as:

- Weather conditions
- Nationality, age, and gender of pilots
- Type, make, and size of planes
- Length of trip
- Port of departure
- Occupancy rate

The number of feasible analyses grows exponentially with the number of variables. So too does the chance of errors and paradoxes.

More data inevitably results in more time spent arguing, validating, reconciling, and replicating. All of these activities create doubt and confusion. There is a real danger that Big Data moves us backward, not forward. It threatens to take science back to the Dark Ages, as bad theories gain ground by gathering bad evidence and drowning out good theories.

Big Data is real, and its impact will be massive. At the very least, we are all consumers of data analyses. We must learn to be smarter consumers. What we need is NUMBER-SENSE.

NUMBERSENSE is the one quality that I desire the most when hiring a data analyst; it separates the truly talented from the merely good. I typically look for three things, the other two being technical ability and business thinking. One can be a coding wizard but lacks any NUMBERSENSE. One can be a master storyteller who can connect the dots but lacks any NUMBERSENSE. NUMBERSENSE is the third dimension.

NUMBERSENSE is that noise in your head when you see bad data or bad analysis. It's the desire and persistence to get close to the truth. It's the wisdom of knowing when to make a U-turn, when to press on, but mostly when to stop. It's the awareness of where you came from, and where you're going. It's gathering clues, and recognizing decoys. The talented ones can find their way from *A* to *Z* with fewer wrong turns. Others struggle and get lost in the maze, possibly never finding *Z*.

Numbersense is difficult to teach in a traditional classroom setting. There are general principles but no cookbook

---

FIGURE P-6 The Flight Delay Data (*Source: The Basic Practice of Statistics*, 5e, David S. Moore, p. 169)

America West

Airport	On Time	Delayed	Delay %
San Francisco	320	129	29%
Seattle	201	61	23%
Los Angeles	694	117	14%
San Diego	383	65	15%
Phoenix	4840	415	8%
Total	*6438*	*787*	*11%*

Alaska Airlines

Airport	On Time	Delayed	Delay %
San Francisco	503	102	17%
Seattle	1841	305	14%
Los Angeles	497	62	11%
San Diego	212	20	9%
Phoenix	221	12	5%
Total	*3274*	*501*	*13%*

Both Airlines

Airport	On Time	Delayed	Delay %
San Francisco	823	231	22%
Seattle	2042	366	15%
Los Angeles	1191	179	13%
San Diego	595	85	13%
Phoenix	5061	427	8%
Total	*9712*	*1288*	*12%*

(see Figure P-6). It cannot be automated. Textbook examples do not transfer to the real world. Lecture materials elevate general concepts by cutting out precisely those elements that would have burned a practitioner's analysis time. The best way to nurture Numbersense is by direct practice or by learning from others.

I wrote this book to help you get started. Each chapter is inspired by a recent news item in which someone made a claim and backed it up with data. I show how I validated these assertions, by asking incisive questions, by checking consistency, by quantitative reasoning, and sometimes, by procuring and analyzing relevant data. Does Groupon's business model make sense? Will a new measure of obesity solve our biggest health crisis? Was Claremont McKenna College a small-time cheat in the school ranking game? Is government inflation and unemployment data trustworthy? How do we evaluate performance in fantasy sports leagues? Do we benefit when businesses personalize marketing tactics by tracking our activities?

Even experts sometimes fall into data traps. If I do so within these pages, the responsibility is solely mine. And if I haven't made the point clear enough, there is never only one way to analyze data. You are encouraged to develop your own point of view. Only by such practice can you hone your NUMBERSENSE.

Welcome to the era of Big Data, and look out!

PART 1

SOCIAL DATA

2 3 4 5 6 7 8

Why Do Law School Deans Send Each Other Junk Mail?

The University of Michigan launched a special admissions program to its law school in September 2008. This Wolverine Scholars Program targeted the top sliver of Michigan undergraduates, those with a 3.80 cumulative grade point average (GPA) or higher at the Ann Arbor campus, allowing them to apply to the ninth-ranked law school as soon as they finish junior year, before the competition opens up to applicants from other universities. Admissions Dean Sarah Zearfoss described the initiative as a "love letter" from the Michigan Law School to its undergraduate division. She hoped this gesture would convince Michigan's brightest young brains to stay in Ann Arbor, rather than draining to other elite law schools.

One aspect of the Wolverine Scholars Program was curious, and immediately stirred much index-finger-wagging in the boisterous law-school blogosphere: The applicants do not have to submit scores from the Law School Admission Test (LSAT), a standard requirement of every applicant to Michigan and most other accredited law schools in the nation. Even more curiously, taking the LSAT is a cause for *disqualification*. Why would Michigan waive the LSAT for this and only

this slice of applicants? The official announcement antici-
pated this question:

> The Law School's in-depth familiarity with Michigan un-
> dergrad curricula and faculty, coupled with significant
> historic data for assessing the potential performance
> of Michigan undergrads at the Law School, will allow us
> to perform an intensive review of the undergraduate
> curriculum of applicants, even beyond the typical close
> scrutiny we devote . . . For this select group of qualified
> applicants, therefore, we will omit our usual require-
> ment that applicants submit an LSAT score.

In an interview with the *Wall Street Journal*, Zearfoss explained the statistical research: "We looked at a lot of historical
data, and [3.80 GPA] is the number we found where, regardless
of what LSAT the person had, they do well in the class." The
admissions staff believed that some Wolverines with exceptional GPAs don't apply to Michigan Law School, deterred by
the stellar LSAT scores of prior matriculating classes.

Many bloggers, themselves professors at rival law schools,
were not eating the dog food. They smelled a brazen attempt
to promote the national ranking—universally referred to as
the *U.S. News* ranking, after *U.S. News & World Report*, the
magazine that has created a lucrative business out of compiling all kinds of rankings—of Michigan's law program. Bill
Henderson, who teaches at University of Indiana, Bloom-
ington, warned readers of the *Legal Profession Blog* that "an
elite law school sets a new low in our obsession of form over
substance—once again, we legal educators are setting a poor
example for our students." The widely followed *Above the
Law* blog was less charitable. In a post titled "Please Stop
the Insanity," the editor complained that "the 'let's pretend
that the LSAT is meaningless so long as you matriculate at
Michigan' game is the worst kind of cynicism." He continued:

"This ploy makes Michigan Law School look far worse than any sandwich-stealing homeless person ever could."

In recent years, *U.S. News* has run a one-horse race when it comes to ranking law schools. By contrast, there are no fewer than six organizations reaching for the wallets of prospective MBA students, such as *Businessweek, The Economist, Wall Street Journal,* and *U.S. News & World Report.* As students, alumni, and society embrace the *U.S. News* rankings, law school administrators shelved their misgivings about the methodology, instead seeking ways to climb up the ladder. Jeffrey Stake, another Indiana University professor who studies law school rankings, lamented that: "The question 'Is this person going to be a good lawyer?' is being displaced by 'Is this person going to help our numbers?'" Administrators fret over meaningless, minor switches in rank from one year to the next. One dean told sociologists Michael Sauder and Wendy Espeland how the university community reacted to a one-slot slippage:

> When we dropped [out of the Top 50], we weren't called fifty-first, we were suddenly in this undifferentiated alphabetized thing called the second tier. So the [local newspaper's] headline is "[School X] Law School Drops to Second Tier." My students have a huge upset: "Why are we a second-tier school? What's happened to make us a second-tier school?"

Schools quickly realized that two components of the *U.S. News* formula—LSAT and undergraduate GPA—dominate all else. That's why the high GPA and no LSAT prerequisites of the Wolverine Scholars Program aroused suspicion among critics. Since the American Bar Association (ABA) requires a "valid and reliable admission test" to admit first-year J.D. (Doctor of Law) students, bloggers speculated that Michigan would get around the rule by using college admission test scores. Several other law schools, including Georgetown University (*U.S. News* rank #14), University of Minnesota (*U.S. News* rank #22), and University of Illinois (*U.S. News* rank #27), have rolled out similar programs aimed at their own undergraduates. At Minnesota, as at Michigan, the admissions officers do not just ignore LSAT scores; they shut the door on applicants who have taken the LSAT.

1. Playing Dean for One Day

Between retaining top students and boosting the school's ranking, one can debate which is the intended beneficiary, and which is the side effect of early admission schemes. One cannot but marvel at the silky manner by which Michigan killed two birds with one stone. Even though the school's announcement focused entirely on the students, the law bloggers promptly sniffed out the policy's unspoken impact on the *U.S. News* ranking. This is a great demonstration of NUM-BERSENSE. They looked beyond the one piece of information fed to them, spotted a hidden agenda, and sought data to investigate an alternative story.

Knowing the mechanism of different types of formulas is the start of knowing how to interpret the numbers. With this in mind, we play Admissions Dean for a day. Not any Admissions Dean but the most cynical, most craven, most calculating Dean of an elite law school. We use every trick in the book, we leave no stones unturned, and we take no prisoners. The *U.S. News* ranking is the elixir of life; nothing else matters to us. It's a dog-eat-dog world: If we don't, our rival will. We are going upstream, so that standing still is rolling backwards.

Over the years, *U.S. News* editors have unveiled the gist of their methodology for ranking law schools. The general steps, common to most ranking procedures, are as follows:

1. Break up the overall rating into component scores.
2. Rate each component, using either survey results or submitted data.

3. Convert the component scores to a common scale, say 0 to 100.
4. Determine the relative importance of each component.
5. Compute the aggregate score as the weighted sum of the scaled component scores.
6. Express the aggregate score in the desired scale. For example, the College Board uses a scale of 200 to 800 for each section of the SAT.

Rankings are by nature subjective things. Steps 1, 2, and 4 reflect opinions of the designers of such formulas. The six business school rankings are not well correlated because their creators incorporate, measure, and emphasize different factors. For example, *Businessweek* bases 90 percent of its ratings on reputation surveys, placing equal weights on a survey of recent graduates and a survey of corporate recruiters while the *Wall Street Journal* considers only one factor, evaluation by corporate recruiters. Note that the scaling in Step 3, known as *standardization*, is needed in order to preserve the required weights applied in Step 5.

Figure 1-1 illustrates the decisions made by *U.S. News* in designing their law school rating. The editors tally up 12 elements, grouped into four categories, using weights they and only they can explain. The two biggest components—assessment scores by peers, and by lawyers and judges—are obtained from surveys while the others make use of data self-reported by the schools.

From the moment the *U.S. News* ranking of law schools appeared in 1987, academics have mercilessly exposed its flaws and decried its arbitrary nature. Since reputations of institutions are built and sustained over decades, it seems silly to publish an annual ranking, particularly one in which schools swap seats frequently, and frequently in the absence of earth-shattering news. Using a relative scale produces the apparently illogical outcome that a school's ranking can move up or down without having done anything differently from

FIGURE 1-1 Components of the *U.S. News* Law School Ranking Formula

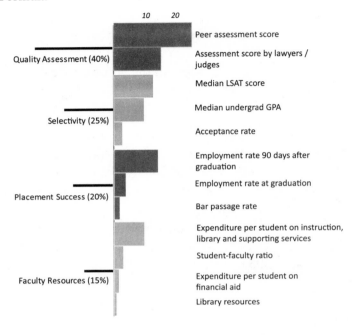

the previous year while other schools implement changes. The design of the surveys is puzzling. Why do they expect the administrators of one school or the partners of one law firm to have panoramic vision of all 200 law schools? The rate of response for the professional survey is low, below 15 percent, and the survey sample is biased as it is derived from the Top Law Firms ranked by none other than *U.S. News*.

Such grumbling is valid. Yet such grumbling is pointless, and has proven futile against the potent marketing machine of *U.S. News*. The law school ranking, indeed any kind of subjective ranking, does not need to be correct; it just has to be believed. Even the much-maligned BCS (Bowl Championship Series) ranking of U.S. college football teams has a clearer path toward acceptance because the methodology can be validated in the postseason, when the top teams face off. The rivalry among law schools does not admit such duels, and thus, we have no means of verifying any method of ranking. There is no such thing as accuracy; the scarce commodity here is *trust*. The difference between the *U.S. News* ranking and the also-rans is the difference between branded, bottled water and tap water. In our time, we have come to adopt all types of rating products with flimsy scientific bases; we don't think twice while citing Nielsen television ratings, Michelin ratings for restaurants, Parker wine ratings, and lately, the Klout Score for online reputation.

The *U.S. News* ranking, if defeated, would yield to another flawed methodology, so law school deans might as well hold their noses. As the devious Admissions Dean, we want to game the system. And our first point of attack is the self-reported statistics. Paradoxically, these "objective" part-scores—such as undergraduate GPA and post-graduation employment rate—tempt manipulation more than the subjective reputation scores. That's because *we* are the single source of data.

2. Fakes, Cherry-Picking, and Missing-Card Tricks

The median undergraduate GPA of admitted students is a signal of a graduate school's quality, and also a key element of the *U.S. News* formula. The *median* is the mid-ranked value that splits a population in half. Michigan Law School's Class of 2013 had a median GPA of 3.73 (roughly equal to an A–), with half the class between 3.73 and 4.00, and the other half below 3.73.

The laziest way to raise the median GPA is to simply fake it. Faking is easy to do, but it is also easily exposed. The individual scores no longer tie to the aggregate statistic. To reduce the risk of detection, we inflate individual data to produce the desired median. The effort required is substantially higher, as we must fix up not just one student's score, but buckets of them. Statisticians call the median a robust statistic because it doesn't get flustered by a few extreme values.

Start with a median GPA of 3.73. If we rescinded an offer to someone with a GPA of 3.75 and gave the spot to a 4.00, the median would not budge, because the one with 3.75 already placed in the top half of the class. So substituting him or her with a 4.00 would not change the face of the median student. What if we swapped a 3.45 with a 4.00? It turns out the median would still remain unaltered. This is by design, as the *U.S. News* editors want to thwart cheating.

Figure 1-2 explains why the median is so level-headed. Removing the bottom block while inserting a new one at the top would shift the middle block down by one spot. The effect

---

FIGURE 1-2 Faking the Median GPA by Altering Individual Data

(a) The median GPA splits the students into two halves. The middle half of students have GPAs in a tight range of 0.27 grade points.

(b) Replacing a 3.75 with a 4.00 does not change the median GPA

(c) Replacing a 3.45 with a 4.00 does not change the median GPA either. But multiple such swaps would work.

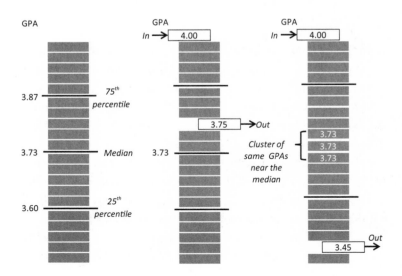

of swapping one student on the median is no larger than the difference between it and the value of its neighbor. This difference is truly minute at an elite program such as Michigan Law School, since the middle half of its class, about 180 students, fit into a super-tight band of 0.28 grade points, thanks to the sieve of its prestige. (For reference, the gap between B+ and A- is 0.33 grade points.)

U.S. News editors might have thought that using the median prevents us from gaming the methodology, but they can't stifle our creativity now, can they? If we swap enough students, the median value will give. Of course, meddling with individual scores is a traceable act. We prefer methods that don't leave crumbs. By obsessively monitoring the median GPA throughout the admissions season, we construct the right profile, student by student, and avoid having to retouch submitted data.

Even more attractive are schemes with built-in protection. Few will condemn us for offering merit-based scholarships to compete with our peer institutions for the brightest students. Financial aid is one of the most important criteria students use to choose between schools. So we divert funds to those applicants with GPAs *just above* our target. At the same time, we withhold scholarships from top-notch students who might prefer our rivals. Instead of awarding one student a full scholarship, why not offer two half-scholarships to affect two applicants?

A flaw of most ranking systems, including the *U.S. News* flavor, is equating a GPA of 3.62 from one school with a GPA of 3.62 from a different school, even though everyone understands each school abides by its own grading culture, teachers create different expectations, courses differ by their level of difficulty, and classmates may be more or less competitive. This flaw is there to be exploited.

We favor those schools that deliver applicants with higher grade point averages. Colleges that take the higher ground—for instance, Princeton University initiated a highbrow "grade deflation" policy in 2004—can stay there while we take the higher GPAs from their blue-collar rivals. Similarly, we like academic departments that are generous with *A*s, and that means more English or Education majors, and fewer Engineering or Science majors. No one can criticize us for accepting students with better qualifications. Cherry-picking schools and curricula occur under the radar, and our conscience is clean since we do not erase or falsify data.

When was the last time you slipped drinks into the movie-plex while the attendant was looking the other way? We play a similar trick on the data analyst. Let's hide (weaker) students. Every year, applicants impress us in many ways other than earning top GPAs. Accepting these candidates sullies our median GPA, and hurts our precious *U.S. News* ranking. Instead of rejecting these promising students, we send them to summer school. Their course load is thus lessened in the fall term, and they turn into "part-time" students, who are ignored by *U.S. News*. Alternatively, or additionally, we encourage these applicants to shape up at a second-tier law school, and reapply after the first year as transfer students, who are also ignored by *U.S. News*.

These tactics exploit *missing values*. Missing data is the blind spot of statisticians. If they are not paying full attention, they lose track of these little details. Even when they notice, many unwittingly sway things our way. Most ranking systems ignore missing values. Reporting low GPAs as "not available" is a magic trick that causes the median GPA to rise. Sometimes, the statisticians attempt to fill in the blanks. *Mean imputation* is a technical phrase that means replacing any missing value with the average of the available values. If we submit a below-average GPA as "unknown," and the analyst converts all blanks into the average GPA, we'd have used a hired gun, wouldn't we? (See how this trick works in Figure 1-3.) If a student suffered depression during school, or studied abroad for a semester where the foreign university does not issue grades, or took on an inhumane course load, or faced whatever other type of unusual challenges, we simply scrub the offensive GPAs, under the guise of "leveling the playing field" for all applicants. Life is unfair even for students at elite colleges; since the same students would have earned much higher GPAs if they had attended an average school, we have grounds to adjust or invalidate their grades. We tell the media that the problem isn't that the numbers drag down our median, but that they are misleading! So good riddance to bad data.

If we let the data analysts fill in the blanks, why not do so ourselves? Our estimate is definitely better since we are the subject-matter experts. Applicants from abroad, for example, frequently have exceptional qualities, but their schools do not use an American-style GPA evaluation system. Instead of submitting "unknown," we exercise our best judgment to award these candidates a grade of 4.00.

We have more drastic options. We can cull the size of our matriculation class. By extending fewer offers of admission, the average offer goes to someone with a higher GPA. Besides,

---

FIGURE 1-3 The Missing-Card Trick: Report the GPAs of "disadvantaged" students as missing. Because of mean imputation, these GPAs are set to the average of the rest of the matriculating students.

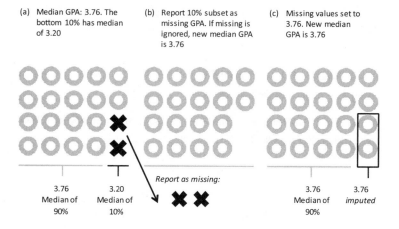

(a) Median GPA: 3.76. The bottom 10% has median of 3.20

(b) Report 10% subset as missing GPA. If missing is ignored, new median GPA is 3.76

(c) Missing values set to 3.76. New median GPA is 3.76

3.76 Median of 90%

3.20 Median of 10%

Report as missing:

3.76 Median of 90%

3.76 imputed

FIGURE 1-4 Downsizing: If the class size is cut, and the pool of applicants remains the same, the GPA scores automatically increase. As word of the school's lower selectivity spreads, it may even attract higher-GPA applicants.

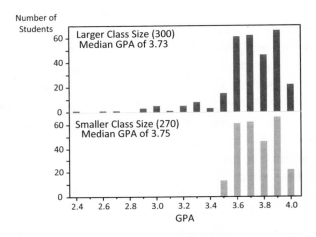

downsizing hikes up exclusivity, and exclusivity attracts better applicants. (See Figure 1-4.) In a sagging economy, we blame the troubled legal profession for the shrinkage. Our finance colleagues may stage a revolt, worrying about foregone revenues—but we'll assure them, every dollar can be recovered, and more, by expanding our second-year transfer program as well as the part-time program.

3. Disappearing Acts, Unlimited Refills, Schools Connect, and Partial Credits

In June 2011, two years after Michigan launched the Wolverine Scholars Program, Dean Sarah Zearfoss felt contented. In a blog post for the school's Career Center, she told students:

> Overall, we've been very happy with our Wolverine Scholar "experiment." I am very optimistic that at the

end of our five-year trial run, we will choose to make it a permanent fixture in our admissions toolkit.

Michigan undergraduates with excellent GPAs have become a special category of applicants who are asked not to submit LSAT scores. This waiver has driven critics bonkers. It seemed like a variant of the missing-card trick, precisely calibrated to nudge the median LSAT score, another component of the *U.S. News* formula.

Most of the tactics we use to manipulate the median GPA carry over to gaming the median LSAT score. Every shift of a below-median score to an above-median score helps a bit. So does dangling scholarship money in front of the right set of students. Enrolling weaker students in part-time programs or "loaning" them to other schools until the second year works just as well. Test takers who are granted "accommodation" status because of documented disabilities such as dyslexia can be removed from consideration. Flunking more first-year students drops those with lower ability from the pool, and as the median LSAT score and GPA elevate, we issue press releases boasting about the toughening of our academic standard.

We contact students with desirable GPAs but unappealing LSAT scores, urging them to re-test. This sure-win tactic deserves ample resources. The LSAT is designed to measure reading and verbal reasoning skills, and has been shown to predict first-year performance at law schools. The Law School Admission Council administers 150,000 tests around the world each year, and everyone who has taken a standardized test knows that one's performance varies with the set of test items, the condition of the testing center, one's mental state on the day of testing, and the relative abilities of other test takers. The LSAT determines ability up to a margin of error, known as a *score band*. LSAT scores, on repeated tests, typically fall into a range of about 6 points on the 120–180 scale. Statisticians consider any score within a score band as statistically equivalent; if they have to choose the best indicator of a candidate's ability, they take the average score. Regardless, we encourage our applicants to submit the maximum score, just like most U.S. schools. The maximum value of anything is likely to be an outlier, and the maximum test score almost surely exaggerates the applicant's aptitude. Students love us for what is in essence "unlimited refills." This policy flushes out the downside of retesting. If the new score is higher, it strengthens the application. If it's lower, the new score melts away. To the Admissions Dean bent on raising the median LSAT score, repeated testing is a godsend. As shown in Figure 1-5, we have weaponized statistical variations: The pool of applicants remains unchanged, and yet, our appreciation of their quality has grown generously.

FIGURE 1-5 Unlimited Refills: For an applicant who takes the LSAT multiple times, the maximum score is never lower than the average or median score. By looking at the maximum score, the entire distribution of scores is shifted upwards. In this example, each applicant is assumed to have taken the test three times.

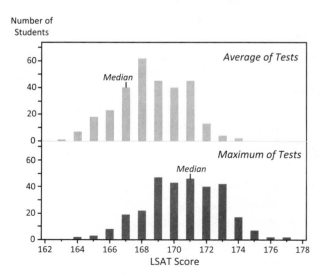

Indeed, every applicant should be required to take the LSAT at least five times. We're getting carried away here—let's start with two test scores, then maybe in a few years, we'll force more re-tests. The only unhappy party is the statistician.

We also want to maintain an impressively low acceptance rate. Ratios are fat targets; we can either reduce the number of admission offers or expand the number of applications. Shrinking the size of each graduating class brings down acceptances; so too does reclassifying weaker students to part-time or transfer status. However, maneuvering the number of offers is constrained by a small class size, say 300 students. With 3,000 applications, our acceptance rate is 20 percent (assuming a yield of 50 percent). Cutting the class size by 10 percent, to 270, moves the acceptance rate to 18 percent. One questions whether this marginal gain is worth a nice chunk of revenues.

Luckily, we can produce the same outcome by finding 334 new applicants—in other words, fixing the bottom part rather than the top part of the ratio. This is chump change for any experienced marketer. Start by waiving the application fee. Then, identify a few segments of applicants with especially low acceptance rates, and advertise heavily to push applications. A delicious example is graduating seniors. Traditionally, professional schools advise students to acquire some work experience before applying. So much for that: We spare no efforts to goad undergraduates to apply, and then only admit the absolute stars amongst them. Outreach to minority groups is another fantastic initiative that boosts our selectivity metric while earning public goodwill.

The most effective plan is sometimes the simplest. We steal an idea that has already spread to almost 500 U.S. colleges: Create a single, unified "Common Application" (Common App). This policy is a major convenience to students. The rationale is the same as why a website encourages new users to bypass the registration process and log on with their existing Facebook or Google Connect credentials. It's also an ingenious way for schools to diminish the acceptance rate, just as for websites to turbocharge the registration rate. With one click, the average student submits the same form to several more schools; the total number of applications explodes. Since none of the participating schools has created any additional first-year spots, the acceptance rate plunges. The mechanism depicted in Figure 1-6 simultaneously applies to every school. It's noteworthy that we produce a symbiosis amid the cutthroat battle for students. The Common App is a tide that lifts all boats.

Having gone this far, we might as well "buy" applicants. You read this right: Pay people to apply. Scores of reputable businesses have exploited this strategy repeatedly. For example, a presence on Facebook has become mandatory for any brand worth its name because hundreds of millions of people hang out in that corner of cyberspace. After Facebook invented the "Like" button, pasting it all over the Internet, marketing managers have seized on it as a metric of success. When CEOs ask the marketing team what they have accomplished, it's not uncommon to get an answer such as, "We got

FIGURE 1-6 Law Schools Connect: When a school receives more applications for a fixed number of spots, the acceptance rate decreases. The Common Application benefits all schools.

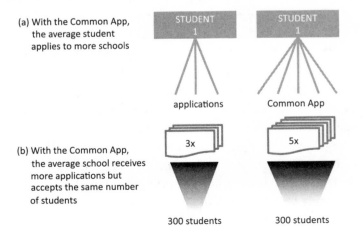

(a) With the Common App, the average student applies to more schools

(b) With the Common App, the average school receives more applications but accepts the same number of students

10,224 Likes through our Facebook promotion this week." Translate this into everyday language: "We told Facebook users we'd send them a free gift if they click on the Like button, and 10,224 of them jumped at it." It takes only a modest budget to entice 334 new applicants.

When the money is tight, we get more creative. Here's another idea: Make sure we count every application, and we do mean, every application, including incomplete submissions, and abandoned applications. (See Figure 1-7.) Separately, double-check each offer before counting it. When someone rejects us, we say that he or she has *voluntarily withdrawn*

FIGURE 1-7 Partial Credits: Since applicants with incomplete forms have a zero percent admission rate, they add to the number of applications, leading to a lower acceptance rate.

(a) Acceptance rate:
20 percent

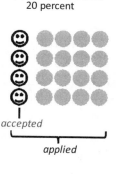

(b) A 10 percent increase in "applications" lowers acceptance rate to 18 percent

(c) Counting enrolled students instead of admitted students also lowers acceptance rate

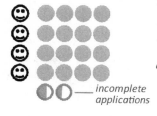

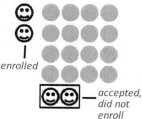

the application. We tally up *enrolled* students, as opposed to *accepted* students. We summon candidates into the office to interrogate them about their first-choice destinations. Why waste an acceptance on a top-drawer applicant who will snub the offer?

4. Creating Job Statistics

U.S. News uses GPA, LSAT scores, and financial resources to measure inputs into education. Outputs are also evaluated, of which job placement is prominent. When our students spend—or borrow—$200,000 to get a law degree, they need high-paying jobs after graduation to justify the investment.

Employment rates follow the same rules as acceptance rates—they are both ratios. We count as many jobs and as few eligible graduates as we can get away with. Amazingly, we can get away with numerical massacre here. Our students self-report their employment status in two surveys, one conducted during the last semester of school and the other within nine months after graduation. We then submit the data to *U.S. News* and other reporting concerns. The National Association for Law Placement instructs us to fill in the blanks before handing over the data. So, to our delight, this game is stacked in our favor.

Missing data is our best friend. The popular mean imputation technique described earlier makes a bold claim: Graduates who ignore the surveys would have provided the same answers as the responders. This claim is false. Those who land at big law firms are more likely to fill out the placement surveys. Graduates who are still unemployed probably won't. Everyone has skin in the game since the *U.S. News* ranking confers bragging rights long after one graduates. To many statisticians, the mean imputation technique is a safe way out. Through it, they avoid having to guess how the others would have responded to those surveys. Since they have not invented any data, it feels as if they are letting the numbers speak. But these numbers mislead, as the hidden, false assumption makes clear. In such cases, another one of which we'll encounter in Chapter 6 on employment data, augmenting the data with reasoned guesses, such as the responders are twice as likely to have landed jobs, should be encouraged. As the devious Admissions Dean, though, we adopt mean imputation precisely because it inflates the employment statistics.

So far, we have shoved the jobless non-responders out of sight, but the employment rate is still tied to the survey data. To loosen that link further, we make another bold claim: A graduate is presumed to have a job unless we unearth evidence to the contrary. This assumption isn't half bad, given the success of previous students in the job market. To proactively gather information, we assign some work-study students to telephone those who haven't answered the placement survey. No, we aren't interested in confirming their jobless status. We call to record voice messages, inviting them to return the call if they want to be counted as unemployed.

As for the second survey, we send only to those who ignore the first one. This environmentally friendly, cost-saving measure ensures that the count of jobs can only go up in the nine months following graduation day. If alumni have lost their jobs in the intervening months, we don't know about it. Taking a page from Uncle Sam (see Chapter 6), we remove graduates who are not actively looking for work, such as those who are taking foreign trips.

We shift our attention from *who* gets counted to *what* gets counted. A job is a job is a job. Not everyone can be an associate in Big Law. We tally up all jobs, part-time as well as full-time, temporary as well as permanent, at big shops as well as at mom-and-pop firms, those requiring Bar passage as well as those that don't. Blending frappuccinos at Starbucks, selling T-shirts at American Apparel, delivering standup comedy at the local bar: These are all legitimate jobs. We call up our friends in high places, courthouses for instance, and arrange for short-term apprenticeships, funded by the law school, of course. In case that's not enough, we hire from within. Our research labs, our libraries, and our dining halls can take extra help. Surely, creating jobs for downtrodden students saddled with unsustainable debt is the morally right thing to do. Let's offer temporary positions to one batch of students at graduation, *before* they fill out the first survey. After six months, we shift the jobs to a second group, in ample time for the second survey.

5. Survey Survival Game, Secret Pacts, and Aided Recall

So far, we have bypassed the two heaviest components of the *U.S. News* ranking. The reputation scores are worth 40 percent of the total. The peer assessment survey is especially influential. Annually, the magazine asks four members of each law school to rate every other school on a scale of 1 to 5 ("marginal" to "outstanding"). The people who have a say include the Admissions Dean, the academic dean, the head of the faculty hiring committee, and the most recently tenured professor. Each member of the quartet votes on any number of schools, although we suspect no one is qualified to pass judgment on all 200 schools. (*U.S. News* does not disclose how many schools the average responder rates—it could be a dozen or a hundred.) The reputation score for each school is averaged over all received votes. This subjective metric is much harder to manipulate than self-reported "objective" data.

About 7 in 10 academic surveys are returned. This rate of response is remarkable, compared to only 12 percent of lawyers and judges for the professional survey. Since most law school deans supposedly detest the *U.S. News* ranking, their eagerness to vote suggests the vast majority is playing the game. We too must pester our four representatives to return the surveys. No occasion is too grand to intervene—not even their kid's first birthday or their new home's closing.

We must assign the bottom rating to all our closest rivals. This isn't arrogance or Machiavellianism, but survival instinct. Consider the inexplicable fact that Harvard and Yale Law Schools, with their towering reputations, accomplished faculties, and distinguished alumni, received average peer ratings of 4.84 out of 5.00 between 1998 and 2008. Apparently, at least 16 percent of those who returned surveys ranked these two schools outside the Top 40 Law Schools in the United States. (This calculation assumes that everyone rated Harvard and Yale in the Top 80.) We owe it to our students and alumni to stay competitive with other deans.

We make secret pacts with mid-ranked schools, especially those on the cusp of breaking into a higher tier. Each side gives the other five stars while we knock a few stars off our respective rivals.

Many observers assume you can't fix survey results. Yes, we can. To assist us in this effort, we hire an authority in brand marketing. The expert tells us these *U.S. News* surveys aren't really about quality of education, but what businesses call *aided brand awareness*. In a typical measurement, consumers are presented with a list of brands and asked which ones they recognize. As expected, those brands with greater recall are more popular. What businesses want even more is *unaided* brand awareness, in which potential customers recall names of brands without hints. It is impossible for any individual taking part in the *U.S. News* surveys to have informed opinions of more than a handful of schools out of the list of 200. But a positive, recognizable brand image can help a school overcome the lack of familiarity.

The branding consultant points out that our promotional efforts only need to reach 800 or so academics, and 1,000 or so lawyers and judges. In reality, an even smaller set of these people are malleable. About 200 questionnaires are returned each year. If we assume each responder votes on 50 schools, then each school's rating represents the opinions of 50 people, on average. Thus, getting even a handful more people to cast a vote makes a difference. Conversely, getting even a handful more people to disparage a rival school also matters.

As the contact information for this audience is by and large public, direct marketing techniques—such as junk mail, spam, and telemarketing calls—are very promising. John Caples's classic book, *Tested Advertising Methods*, contains a wealth of best practices accumulated over many decades of scientific testing. Successful headlines appeal to self-interest or convey news. Long copy that is crammed with sales arguments beats short copy that says nothing. Keywords such as *Announcing*, *New*, and *At last* produce results. Avoid poetry or pompous words. Repeated communications reinforce the marketing message. Glossy materials stand out from the stack of junk mail. These, and other learning, have been carefully tested.

Typically, the marketer creates two versions of a message and compares the number of responses to each version. For example, one mailing leads with *"Announcing* a great new car,"* while another reads "A great *new* car." When the two groups of recipients are made as similar as possible, the comparison is valid. If money is no object, we flood the marketing materials to a wider audience, such as academics outside the dean's offices and legal professionals of all stripes. Since these people circulate in some of the same social networks as our targets, we benefit from a "halo effect."

No subjective metric can escape strategic gaming (I'll return to this in Chapter 2). Every factor used by *U.S. News* can be manipulated. The possibility of mischief is bottomless. Fighting ratings is fruitless, as they satisfy a very human need. If one scheme is beaten down, another will take its place and wear its flaws. Big Data just deepens the danger. The more complex the rating formulas, the more numerous the opportunities there are to dress up the numbers. The larger the data sets, the harder it is to audit them. Having NUMBERSENSE means:

- Not taking published data at face value
- Knowing which questions to ask
- Having a nose for doctored statistics

Perhaps you're wondering: Instead of NUMBERSENSE, can consumers of data count on decency and integrity?

6. Guilt by Association

In November 2011, the *Above the Law* blog landed the final blow in its tussle with Sarah Zearfoss, admissions dean of Michigan Law School. The blogger noticed the quiet demise of the Wolverine Scholars Program. A gander at the Michigan Career Center blog revealed a new preamble to Zearfoss's midterm appraisal of the special admissions policy. It advised readers that the program was scrapped in July—not quite a month after Zearfoss had extolled its virtues.

The blogger found out about the U-turn from an interview Zearfoss gave to the *Daily Illini*, the student newspaper of the University of Illinois. Zearfoss said, "The program was not producing the results the school had originally hoped for, and thus, was discontinued." She did not explain the change of hearts. The central subject of the *Daily Illini* piece was Zearfoss's counterpart at the University of Illinois College of Law (*U.S. News* #21), Paul Pless, who, in 2008, launched iLEAP, a special admissions program for Illinois undergraduates similar to Michigan's.

Identified as "a maverick and a reformer," Pless trumpeted the brilliance of his invention:

> [With iLEAP,] I can trap about 20 of the little bastards with high GPAs that count and no LSAT score to count against my median. It is quite ingenious. And I thought of it before Michigan, they just released it earlier. I was hoping to fly under the radar.

The correspondent complimented Pless, saying first "that is clever," and later "nice gaming the system, I'm so proud." The admiration prompted Pless to describe a further aspect of the plan: "if I don't make [the applicants] give me their final transcript until after they start, I report the GPA that was on their application." Pless was worried, as he should, that the rising seniors, who have secured law school spots in the fall, might take their feet off the pedals. That GPA on their application, of course, has an artificial floor, just as it did at Michigan. The *Daily Illini* learned that the average GPAs of iLEAP classes have exceeded 3.80. It appears that guilt by association was too much to bear for Pless's peers at Michigan.

The unsightly e-mail exchange came to light in November 2011, when the Illinois College of Law (COL) confessed to committing massive reporting fraud for at least six years. Under Pless's leadership, the Admissions Office submitted falsified data to *U.S. News*, and other reporting agencies. In 2011, they inflated the undergraduate GPA from 3.70 to 3.81, large enough to necessitate altering almost one-third of the individual GPAs. In addition, eight international students who did not have GPAs as well as 13 others admitted under iLEAP were assigned 4.00 against the rules. In 2009, the acceptance rate was reported as 29 percent, unchanged from 2008, when in reality, Illinois gave offers to 37 percent of applicants. Admissions offers were undercounted, after inappropriately removing students who "withdrew before deposit." Applications were overcounted, by including candidates for transfers and advanced study who were not part of the J.D. program.

Between 2006 and 2011, Illinois also lifted median LSAT scores from 163 to 168. The impact of such progress was not lost on Pless, who contributed the following comments to the 2006 Strategic Plan for COL:

> The three-point LSAT median increase [from 163 to 166] that we accomplished in the last year alone is, as far as we know, unprecedented in the history of the legal academy . . . . Because the *U.S. News* law school rankings place so much weight on student credentials, COL would have moved from 27th to 20th in last year's

rankings had we been able to report this improvement a year ago (holding all else constant).

In its 2008 Annual Report, two years hence, COL discussed another gambit to skew the median LSAT score that was stuck at 166. The school had dramatically expanded the disbursement of scholarships, fourfold in four years. The financial aid came in the form of tuition remission, with a median grant of $12,500 in 2010. But the staff warned that returns would be diminishing. "To move from 166 to 167 would in our estimation take over a million dollars in new scholarship money," they stated. The school also intended to "drastically raise tuition" and "funnel a lot of that back into scholarships, both to reduce the burden on our students and to increase our spending for *U.S. News*." In 2011, when every single student, including those taken off the front of the wait list, received at least $2,500 in aid, Pless delivered a miracle, a median LSAT score of 168. It emerged that the real number was only 163, and to bolster it by 5 points, he doctored the scores of 60 percent of the class. It took a huge hammer to pound the median into submission.

The actions by Pless's office cannot be labeled rogue. As noted in the investigative report commissioned by the school in the wake of this scandal, COL set aggressive targets for the median LSAT score and the median GPA for each upcoming J.D. class. The five-year plan (2006–2011) created targets of 168 and 3.70. Pless recruited faculty members to simulate the rankings under different combinations of LSAT score and GPA. In an e-mail from early 2009, he told the Dean, "the Lawless calculator projects a 4-place improvement with the 165/3.8 over the 166/3.7." (What an unfortunate name! Robert Lawless, a professor at Illinois, developed a way to predict *U.S. News* rankings.) Later that year, the Dean informed the Board of Visitors, "I told Paul we should push the envelope, think outside the box, take some risk, and do things differently." Over the years, Pless was showered with praise and paid for his consistent ability to deliver the goods.

In February 2011, Villanova Law School (*U.S. News* rank #67) admitted some of the data used by *U.S. News* was "inaccurate." In a series of memos issued to alumni, the Dean disclosed that GPAs and LSAT scores were inflated for five years, and the number of admissions offers was "inaccurate" the past three years. The school congratulated itself for conducting "a textbook investigation... prompt and comprehensive," and for "expanding the investigation" ... "on our own initiative." However, unlike Illinois, Villanova did not come clean on the extent and the methods of the ratings scam. The *Philadelphia Inquirer* shamed this "unseemly silence," and their refusal to release the investigative report.

In July 2005, the *New York Times* detailed how Rutgers School of Law, Camden (*U.S. News* rank #72) sought to scale the ranking table by expanding its part-time program. Summer classes were held for those with lower LSAT scores or GPAs so that they did not qualify as full-time students in the fall term when *U.S. News* collected data. Rutgers-Camden's full-time enrollment has fallen for seven consecutive years. Dean Rayman Solomon told the reporter: "There's an educational benefit, a financial benefit, and a residual *U.S. News* benefit." Baylor University's School of Law (*U.S. News* rank #50) benefited from a similar policy.

7. Law Schools Escaped the Recession

In May 2010, Paul Caron, a law professor at the University of Cincinnati, posted a startling chart on his *TaxProf Blog*, showing a steep upward line from 35 percent to almost 75 percent between the years 2002 and 2011. As the U.S. economy tanked, evidently more and more law schools no longer knew what their students were doing immediately after graduation. By 2011, three out of four law schools failed to submit this data to *U.S. News*. They therefore acquiesced to the magazine's mysterious, but publicly announced, formula to fill in the blanks: *Employment rate at graduation is taken to be roughly 30 percent lower than the employment rate 90 days after graduation,* a number that almost all schools continue to supply, perhaps because it is an American Bar Association (ABA) requirement. Of the 200-odd accredited schools ranked by *U.S. News*, Caron found only 16 schools to have self-reported employment rates at graduation that were 30 percent or more below the rates after 90 days. Several of these schools could have gained appreciably in the rankings if they had just withheld the data. Every one of the honest 16 was ranked Top 80 or below, with the majority in Tier 3 (100–150 out of 200). No school in the top half of the table gave *U.S. News* an employment rate lower than what the editors would have imputed. Incredibly, the *U.S. News* editors responded to Caron's discussion by announcing they would henceforth change the method of imputation and withhold the revised formula from the public. Hiding information will not stop enterprising law school deans from reverse-engineering the formula; nor would it deter manipulation.

Astute readers of Caron's blog noticed that those 16 schools, mostly ranked outside the Top 100, claimed that 89 to 97 percent of their students found jobs within 90 days of graduation. Indeed, *U.S. News* told its readers over 90 percent of graduates found jobs within nine months in four out of ten law schools that are good enough to be ranked by the magazine in 2011. At nine schools, 97 percent or more found work. University of Southern California (*U.S. News* rank #18) reported, with a straight face, an employment rate at nine months of 99.3 percent, putting the top programs like Yale, Harvard, and Stanford to shame. Imagine you were the only one in the 200-strong Class of 2009 to remain jobless! Against these statistics, two Emory law professors evoked a reality that few people in the trenches could deny: "Since 2008, the legal profession has been mired in the worst employment recession—many would argue it is a depression—in at least a generation."

In April 2012, ABA released details of employment for newly-minted J.D.s. For the first time ever, accredited schools broke down the jobs into categories, such as temporary or permanent positions, and whether the positions are funded by the schools themselves. ABA revamped the reporting guidelines under pressure from critics who guffaw at the dreamy employment rates that are turned in by law schools year after year, and gobbled up by *U.S. News* editors unsalted. The ABA data dump, assuming it could be trusted, revealed that only 55 percent of the so-called employed have full-time, long-term jobs requiring a J.D. The majority of the accredited law schools performed even worse than that level. Many of the jobs, especially those counted by lower-tier schools, do not pay enough to cover the student loans. Besides, a quarter of the schools created jobs for 5 percent or more of their graduating classes. Higher-ranked schools tended to be more eager job makers: Yale University (*U.S. News* rank #1), University of Chicago (*U.S. News* rank #5), New York University (*U.S. News* rank #6), University of Virginia (*U.S. News* rank #7), Georgetown University (*U.S. News* rank #13), and Cornell University (*U.S. News* rank #14) all featured in the top 10 percent, hiring between 11 and 23 percent of their own graduates. Since 2010, Southern Methodist University (SMU) Dedman School of Law (*U.S. News* rank #48) has paid law firms to hire its graduates for a "test drive," basically two-month-long positions. About 20 percent of the class participates in this program. SMU considers these jobs funded by employers, even though they pay nothing out of pocket.

Beyond such inconceivable employment rates, the law schools delivered another remarkable feat by supplying placement data for 96 percent of all graduates. That rate of response is unheard of in any kind of surveys. Writing for the *Inside the Law School Scam* blog, Paul Campos, a law professor at the University of Colorado, Boulder, found that one in ten of those with missing data came from a single school, Thomas M. Cooley Law School (*U.S. News* Tier 4). Cooley's website sheds light on how the ABA allows law schools to invent job statistics. Every graduate is presumed to have full-time, long-term employment unless contradicted by evidence. Richard Matasar, Dean of New York Law School, once wrote about several "legendary" . . . "tricks of the trade" in the ratings game. One tactic involves "calling graduates, and leaving them messages that if they do not call back, you will assume that they are employed." We also learn from Cooley's disclosure that someone working for a legal temp agency is considered to be employed full-time and long-term.

In May 2012, Hastings College of the Law (*U.S. News* rank #44), a part of the University of California system, declared a plan to cut enrollment by 20 percent over a three-year period. Dean Frank Wu explained some of the benefits of this austere measure: "As a smaller school, we will have better metrics. Students will have a better experience, and obviously there will be better employment outcomes." A rise up the ranking table is an expected result. In response to suspicion in some quarters, the Dean issued the following statement:

> UC Hastings takes rankings seriously and intends to do everything we can to improve ours, and we've shown our ability to analyze the statistics and then take action; however, we will do only what is academically beneficial and ethical.

Within months of Hastings's announcement, George Washington University also said it would reduce the class size of its law school (*U.S. News* rank #20). Others will no doubt follow suit.

8. Sextonism

In August 2005, Brian Leiter, a law professor at the University of Chicago who publishes an alternate ranking of law schools based on his own criteria, started a "Sextonism Watch" on his blog, *Brian Leiter's Law School Reports*. John Sexton is a former dean of the School of Law, and the current president of New York University. Among law faculty, Sexton is credited with inventing *law porn*, which is basically junk mail containing "uncontrolled and utterly laughable hyperbole in describing its faculty and accomplishments to its professional peers" sent to thousands of law school staff across the country. One of NYU's earliest marketing efforts was a glossy magazine-cum-brochure with a color photograph of celebrity philosopher-lawyer Ronald Dworkin on the front cover and the aggrandizing title, *The Law School*, with the article set above the compound noun. Almost six pounds of junk mail—totaling 43 pieces, including eight glossy brochures—arrived within a single week in the mailbox of another law professor blogging anonymously at *The Columnist Manifesto*. Jim Chen, who taught at the University of Minnesota at the time and writes for the *MoneyLaw* blog, saw it differently, defining *Sextonism* as "the adroit (if not altogether credible) promotion of an educational institution among its constituents and its rivals alike."

Since 2005, many other schools have joined the scramble for "mind share." Decades of consumer research leave little doubt that direct marketing enhances aided brand recall, which can influence law school deans or lawyers who fill out *U.S. News* surveys. The professional quality of the promotional materials suggests that law schools have set up sophisticated branding operations. They are testing a variety of formats, papers, and designs, just like mature businesses. They are utilizing gifts and offers to vie for attention, just like experienced advertisers. At the University of Alabama School of Law, Paul Horwitz, with the help from readers of his blog, *PrawfsBlawg*, cataloged the bounty of vanity stuff given out to visiting professors: coffee mugs, hats, knitted caps, notebooks, bags, kitchen magnets, coasters, clocks, book lights, chocolates, wine, coffee beans, and so on, all embossed with school logos. In marketing parlance, these "high-impact pieces" are expected to rise above the glut.

9. The Steroids Didn't Help

By the 2000s, there is little doubt that our devious Admissions Dean has leapt from the pages of fiction to the august offices of law schools around the country. A succession of scandals threatens the authority of the *U.S. News* ranking, and erodes the credibility of school administrators. Institutions entrusted with educating the next generation are caught with their pants down, engaging in unethical practices. The educational benefits of these policies are at best dubious, and at worst duplicitous. While some offenses, such as the audacious doctoring of over half of the LSAT data, probably are not widespread, other tactics, such as inventing job statistics and reclassifying full-time students as part-time, are considered tools of the trade. You can almost hear the Lance Armstrong apologists, arguing that it's not cheating when "everyone" else is doing it.

What we've witnessed is clearly the tip of the iceberg. In addition to the above, researchers also noticed a dramatic spike in:

- The attrition rate of first-year law students
- The inappropriate booking of phantom expenses to boost per-student expenditures
- The possible overcounting of graduates landing jobs at major law firms

Meanwhile, the cheating scandal engulfed the famous college rankings of *U.S. News*. Claremont McKenna College (*U.S. News* rank #9 in national liberal arts colleges), Emory University (*U.S. News* rank #20 in national universities), and Iona College (*U.S. News* rank #30 in regional universities— north) each admitted to manipulating a broad array of statistics. The Naval Academy was accused of including incomplete applications to sustain the myth of its extraordinarily low admission rate. Several colleges in New Jersey were found to have inflated SAT scores.

American universities have become vast bureaucracies incapable of reforming from within. In each of these scandals, the top administrators interpreted their role as damage control and public relations management, rather than cultural change and ethical renewal. The investigators, hired by the dean of the college, blamed one lone ranger or a few bad apples in the admissions office. Every administration excused itself.

The University of Illinois College of Law (COL) blamed "a single employee ... for this data manipulation." The investigative report from Illinois stated without irony: "COL and its administration, under the leadership of the current Dean, are appropriately committed to the principles of integrity, ethics, and transparency, and communicate this commitment with appropriate clarity and regularity." The Dean was misunderstood when he called for "pushing the envelope" and "thinking outside the box."

The "unseemly silence" at Villanova Law School did not prevent the administrators from disclosing that "individuals [in the admissions office] acted in secret ... neither the Law School nor the University had directly or indirectly created incentives for any person to misreport data."

The President of Claremont McKenna College was "gratified that the [investigative] report confirmed that ... no other employee [but the Admissions Dean] was involved. . . . This was an exceptional incident."

The staff at these institutions exonerated each other by arguing that their obsession with the *U.S. News* ranking, their setting of target GPA and LSAT scores, the use of spreadsheets to predict ranking shifts under hypothetical scenarios, and the celebration and reward structure for reaching those goals, conform to the industry norm. Why the same standard of judgment by peers does not protect admissions staffs from condemnation was never explained.

And then, the prestigious Claremont McKenna College (CMC) in Southern California pulled out the taking-steroids-did-not-help excuse. From 2004 to 2012, the school submitted falsified data on average and median SAT scores, average and median ACT scores, distribution of SAT section scores, proportion of students graduating in the Top 10 percent of their high-school class, and admission rate. According to the *Los Angeles Times*, Pamela Gann, President of CMC, re-marked: "The collective score averages often were hyped by about 10 to 20 points in sections of the SAT tests . . . That is not a large increase, considering that the maximum score for each section is 800 points."

Not a large increase? Is the administration willfully igno-rant, or just ignorant? The reporter dutifully printed Gann's comments, without comment. If he had NUMBERSENSE, he should have realized that 800 was a red herring. Adding 10 or 20 points to an individual's score would have been more like a hiccup than whooping cough, but adding 10 or 20 points to the *average* score is pneumonia; it is fraud of the gravest scale. This is equivalent to boosting the individual scores of about 300 freshmen by 10 or 20 points *each*, totaling 3,000 to 6,000 phantom points! Now double that, as there are two sections, Verbal and Math.

The investigators discovered that CMC embellished the average combined SAT score by 30 to 60 points, depending on the year. (Gann chopped this in half, rounded down, and reported a modification per section.) It's true that the maxi-mum combined score is 1,600. Stop for a moment, and think what it means for the average score to be 1,600. It means every one of about 300 individual scores has to be 1,600. What a dumbfounding distraction. We should instead be pay-ing attention to how much the average combined score varies from year to year. Statisticians use the *standard error* to de-scribe this variability: here, it's 10 points. (This is illustrated in Figure 1-8.) A simple way to understand the standard error is that two-thirds of the time, the average score falls into a narrow 20-point band. A 30- or 60-point inflation of the average score, therefore, is an outrage. This fraud is be-tween three and six standard errors when a deviation of three standard errors from the norm is regarded as extreme. Take any normal year in which the average score is at the 50th percentile of the historical range. A 30-point shift takes the number up to the 99.7th percentile. It's like upgrading every C student to an A. To label this manipulation "not large" is simply embarrassing.

The calculation, in fact, understates the scale of the deception because the spread of the average scores is smaller than the 20-point band. This bandwidth assumes that the

FIGURE 1-8 Doping Does Not Help, So They Say: The President of Claremont McKenna College compared 10 to 20 points of augmentation to the full range of 800 points, while a proper analysis should compare 30 to 60 points to the normal spread of 20 points.

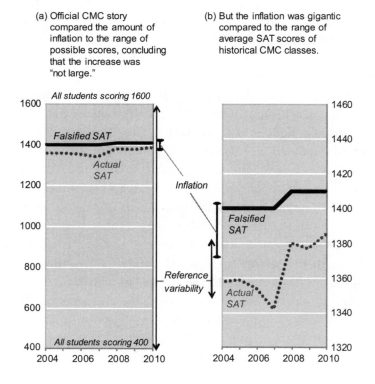

(a) Official CMC story compared the amount of inflation to the range of possible scores, concluding that the increase was "not large."

(b) But the inflation was gigantic compared to the range of average SAT scores of historical CMC classes.

matriculating students form a random sample of all SAT test takers. But surely students at the nation's ninth-ranked liberal arts college are concentrated at the upper end of the SAT scale.

This is a quintessential moment that calls for NUMBERSENSE. When a college president, or other respected person, throws out a statistic, we must not rely on blind trust. NUMBERSENSE is that bit of skepticism, urge to probe, and desire to verify. It's having the truffle hog's nose to hunt the delicacies. Developing NUMBERSENSE takes training and patience. It is essential to know a few basic statistical concepts. Understanding the nature of means, medians, and percentile ranks is important. Breaking down ratios into components facilitates clear thinking. Ratios can also be interpreted as weighted averages, with those weights arranged by rules of inclusion and exclusion. Missing data must be carefully vetted, especially when they are substituted with statistical estimates. Blatant fraud, while difficult to detect, is often exposed by inconsistency.

Can a New Statistic Make Us Less Fat?

In front of you are five sachets. Four of them contain milk-shake powder, three chocolates, and one banana. They are soluble in cold water, and remind one of Nesquik instant chocolate milk. The other packet is a chicken soup powder, which you'll dissolve in hot water. You are looking at the entirety of your day's ration. Yes, four cold shakes and a hot soup are all you're permitted to consume. That, and eight glasses of water. All liquids. Combined, they supply 800 calories, about 14 grams of proteins, 20 grams of carbohydrates, and 3 grams of fat. The time is 8 a.m., and you're about to make your first shake. You'll drink a glass of water, perhaps two. For lunch, you'll have another shake. Three hours later, you'll have another one. The soup is your full dinner, and the banana shake is your nightcap. Every day, for a minimum of 100 days, this is your ration.

You are not to be sedentary. Physical activity is mandatory five times a week, 60 minutes at a time. Because of your liquid meals, your head may feel light, and your legs may tire. Initially, you may last only 20 minutes before falling down from exhaustion. In the event of such embarrassment, you must come clean at the next confessional. You report once a week to your handler, who records your every aberration. Every two weeks, your vitals are to be inspected, and your progress monitored.

Who are you? Obviously, you are not a normal adult, who needs 2,000 to 3,000 calories a day. You have given up some degree of freedom, not just in your choice of food but also in your daily activities. You allow another person to dictate key aspects of your life. You have a strong character and admirable willpower. You don't bend easily. Your threshold for pain is high.

You are a dieter on the OPTIFAST program. Novartis Nutrition Company, now part of the conglomerate Nestlé, created OPTIFAST in 1974, and it truly became a household name in November 1988 when Oprah Winfrey snuggled her new body into size 10 Calvin Klein jeans on her mega-popular talk show, crediting the diet for a 67-pound weight loss in four months. Over a million people are said to have tried the diet. Typically, about half of them persevere to the end, by which time solid foods have been gradually reintroduced. In addition to stamina, patients put their money where their mouths are. A standard treatment program lasting 18 weeks costs up to $3,000.

1. The Achilles' Heel

Oprah's celebration did not last. She regained 10 pounds in just two weeks after going off the diet. Within four years, she would hit her highest-ever weight of 237 pounds. And she isn't alone. The Institute of Medicine of the National Academy of Sciences found that 98 percent of people on diets returned to their original weight within five years. This is the Achilles' heel of every diet known to humans. It is much, much harder to maintain your weight than to lose it.

"Why can't I solve this problem?" asked Darrell Phillipson, a retired judge of King County, Washington. He lamented: "It's not a question of willpower; it's not a question of brain power." The 63-year-old has been fighting his weight for over 40 years. He biked. He hiked. He worked out. He joined Overeaters Anonymous. He adopted a low-carb diet. He tried Weight Watchers. He also went on OPTIFAST. Phillipson was a yo-yo dieter, just like Oprah. Nothing he sowed reaped. When he retired in 2011, he carried 425 pounds.

Judge Phillipson's story is featured in *The Weight of the Nation*, an ambitious, four-part documentary film by HBO about obesity in America. The 2012 film paints a bleak reality. We meet many flailing dieters: one named Audrey lost 30 to 50 pounds 50 or 60 times in her life. Exercise, the primary tool of intervention prescribed by doctors everywhere, is shockingly ineffective—one candy bar needs 30 minutes of running to work off; one slice of pizza, an hour; and one regular-sized hamburger, three hours and 15 minutes. Experts complain that the NBC reality show *The Biggest Loser* sends the wrong message, because physical exertion simply cannot produce such rapid weight loss. A twin brother who is obese promised the other who isn't to usurp his genes, knowing that 60 to 80 percent of the risks of obesity are inherited. Small-time Iowan farmers protest the lunacy of a federal farm policy that subsidizes the planting of corn and soybeans by agribusiness behemoths at the expense of fruits and vegetables, which occupy less than 3 percent of farmland. The soft-drinks industry made a deal with President Bill Clinton to pull sodas from elementary schools, but nothing really changed as the vendors just replaced Coke and Pepsi with fruit juices and sports drinks produced by the same multinationals, both of which deliver an equal amount of "empty" calories, calories that have no nutritional value as they come strictly from sugars. One school of thought holds that a short-term weight drop triggers the body to defend its canonical weight, making it tough for dieters to settle at their new weight.

The power of *The Weight of the Nation* comes from the accumulation of individual failures. Millions of individual failures stack up to a national crisis. Consider this analogy. What if the Department of Education created a new mathematics test for ninth graders, and set a target that by the fifth year, every district should have at least 30 percent of students achieve a passing grade? We should be horribly embarrassed if not one district reached this modest goal. (And what if the standard of passing has also been lowered to getting a third of the questions correct, not half as we expected? We should cower in shame. This in fact happened in 2008 with New York State, while it was earning "No Child Left Behind" accolades. I will deal with such shenanigans later.)

When the Centers for Disease Control (CDC) launched the "Healthy People 2010" campaign, the prevalence of adult obesity did not exceed 30 percent in any state. CDC challenged every state to bring the rate to below 15 percent within a decade. How many states hit this target? Not one. The least obese state, Colorado, went shy of the goal by 6 percent. At the close of the campaign, 12 states breached the startling 30 percent level.

Obesity began its alarming ascent in America in the 1980s. All of a sudden, a greater and greater proportion of adults became obese. While the prevalence remained flat around 14 percent for two decades, by the early 1990s, it jumped to 21 percent for men and 26 percent for women; these numbers continued rising so that by 2000, 28 percent of men and 34 percent of women were obese. By 2008, the statistic for men reached 32 percent, while that for women inched up to 35 percent. This obdurate winless streak leaves healthcare experts scratching their heads. Why can't they solve the problem?

2. The Baloney Mass Index

Diets are short-term fixations; even the effective ones lead to only a 10 percent reduction in body weight, and almost every pound lost is regained soon enough. Food portions keep getting larger. Physical exercise doesn't burn off the fat quickly enough. Dr. James Hebert, an epidemiologist at the University of South Carolina, conceded, "We're stuck right now," wearing the same haplessness that shadows the HBO narrative. The Gamecocks' home state is one of 12 states where more than 30 percent of adults are obese. The way out, Dr. Hebert suggested, is a better way to "measure the [obesity] problem and the public health consequences."

This theme of better measurement reverberated in the mass media around the same time as the HBO documentary, thanks to new research by Dr. Nirav Shah, the New York State Commissioner of Health, and Dr. Eric Braverman, who runs a wellness clinic in Manhattan. Drs. Shah and Braverman touted an alternative way of defining obesity. They warned that the average American is much fatter than previously acknowledged. A more accurate metric would improve public policy and medical treatment. If true, the desperate fight against obesity can be won by changing how we define fatness. It's all very enticing, this easy-win scenario. The press soaked all this up. When the media features a research study, it is legitimized and amplified. How much should you trust this research? What does "much fatter than acknowledged" mean?

The enemy of Drs. Shah and Braverman is the ubiquitous Body Mass Index (BMI). BMI is how your family doctor judges if you're too fat. It is the metric used by the National Institutes of Health (NIH) in its reporting. To no one's surprise, BMI is introduced from the first minute of the first part of *The Weight of the Nation*, and then it grows roots through all four episodes; how entrenched it is in the language and thinking of the experts. Viewers hear hardly any doubts about the obesity metric, computed by dividing one's weight (in kilograms) by one's height (in meters) twice.

Ancel Keys was a professor of physiology at the University of Minnesota who coined the term Body Mass Index. Keys is remembered primarily for linking saturated fats to cholesterol to heart disease. The Mediterranean diets that sprang from his theory have undergone a revival in recent years. Keys was always the interventionist who believed governments ought to expediently promote preventive health care. His 1972 paper that launched BMI as a global standard is mostly forgotten today. It wasn't Keys who discovered the inverse-square formula, though.

That weight is proportional to the square of one's height was noticed in the 1830s by Belgian scientist Adolphe Quetelet, one of the first statisticians to insert mathematics into the social sciences. At the time, Quetelet was molding his groundbreaking creation, the "average man," and searching for universal constants relating weight to height. He observed that individuals of an age group conformed to such constants. He believed like people should have like weight-to-height ratios. In modern times, the medical profession picked the range of 18.5 to 25 to be the ideal BMI; and we worry about deviants.

The usual figures used to size up the obesity epidemic come from NIH, which tells us that 34 percent of American adults are obese. This number is scientifically derived from the 2008 National Health and Nutrition Examination Survey (NHANES), from interviews and physical exams of a representative sample of 10,000 people each year. The definition of obese is having a BMI greater than 30 (BMI > 30). We can get a mental picture of this metric: a five-foot-five woman is fat if she weighs more than 180 pounds; likewise a six-foot-two man is fat if the scale tops 234 pounds.

Alternatively, Drs. Shah and Braverman measure the percent of body fat directly. By their count, the proportion of obese Americans ought to be 64 percent, not 34 percent. If they are right, the BMI metric has a deplorable level of inaccuracy. Dr. Braverman and his associates assembled and analyzed the records of 1,400 patients who took a dual-energy X-ray absorptiometry (DXA) scan at their clinic, PATH Medical. Originally developed to diagnose possible osteoporosis, DXA produces a profile of body composition, split into bone, muscle, and fat. By contrast, BMI can't distinguish between fat and muscle, both of which contribute to one's weight while only fat is thought to signal early death. (PATH Medical, which funded Dr. Shah's research, is unusual in having 18 percent of their patients take DXA scans during their first visits, and 71 percent within three weeks.) The doctors discovered that BMI, as compared to DXA, incorrectly classifies 40 percent of the patients; and almost all the mistakes take the same form: the BMI > 30 cut-point fails to flag some DXA-obese people.

Drs. Shah and Braverman sell us a ray of hope. Drop the "baloney mass index" and adopt DXA, and all will be well.

3. The Perversion of Measurement

Who is obese is up to one's definition. It's a quantity with no objective value, leaving room for negotiation and manipulation. In other words, it's just like any metric out there. Look at teacher quality, student aptitude, employee performance, wine rating, customer satisfaction, and business profitability. Since not one of these has an intrinsic value, it's anyone's guess what "accuracy" means.

One thing is for sure. Measuring anything subjective always prompts perverse behavior. The State of New York shamelessly reduced the passing standard in standardized tests so as to beat "No Child Left Behind" targets. As the pay-for-performance movement in education gains ground, more and more below-average students are pushed to quit, lest they drag down the schools' test scores. Groupon, the daily deals company which I'll discuss in Chapter 4, got into trouble with the Securities and Exchange Commission (SEC) for inventing its own profitability metric named "Adjusted Consolidated Segment Operating Income." Do you recall the last time a customer service agent treated you unusually well? Was there a customer satisfaction survey waiting for your response soon after?

All measurement systems are subject to abuse. The debate over obesity metrics provides a great vantage point to explore how easy it is to lose one's bearing. Here is the NUM-BERSENSE guide to following this controversy.

a. When Outcomes Disappoint, Change How They Are Measured

Failure is hard to swallow. Faced with disappointment, people often ask what's wrong with the metric, rather than what's wrong with the program. Time spent negotiating how to tweak the metric takes time away from finding new ways to improve the outcome.

The *Los Angeles Times* reporter, Melissa Healy, observed that "in the last two years, researchers . . . are using [a wide range of alternatives to the BMI] to measure the effectiveness of interventions such as weight-loss counseling, exercise regimens, and drug therapies." She implied that such treatments have failed in the past but only because they failed to affect the Body Mass Index. Given better measures, these treatments would magically become effective.

The NIH estimates that 36 percent of adult women are obese, based on the BMI > 30 cut-point. Seventy-four percent of Dr. Braverman's female patients are obese by the DXA metric. This 38-point gap represents people who are DXA-obese but not BMI-obese. On further inspection, NIH has a name for this group: *overweight*. Thirty-nine percent of adult women are deemed overweight, with a BMI between 25 and 30. So adopting DXA, in effect, merges the overweight and obese categories into one.

Would reclassifying overweight people as obese stall the obesity epidemic? I wouldn't bet on it. As shown in Figure 2-1, the relationship between body size and mortality is not linear. The best current research suggests that it is a U- or J-shaped curve. While the obese (BMI > 30) and the underweight (BMI < 25) can expect marginally higher risks of death, people with a BMI between 25 and 30 tend to live longer than average. The proposed relabeling may prescribe nothing more than unnecessary treatment.

FIGURE 2-1 The Curved Relationship between Body Mass Index and Mortality: Overweight people may outlive both the obese and the underweight.

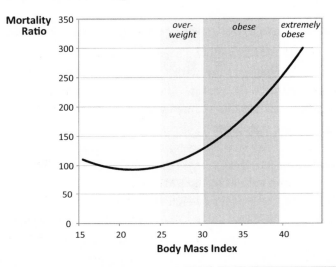

There is also a sizable group who are obese by both metrics. If current policies and diets are ineffective, the same policies and diets would still fail them even after the metric is switched.

b. The More Metrics Change, the More They Stay the Same

Data analysts know well most metrics are strongly correlated. After all, they are supposed to measure the same thing.

Researchers from Columbia University, the University of Cambridge, and Jikei University (Japan) found that BMI and percent body fat have a correlation between 0.7 and 0.9 in all three countries; that's *very* high. Advocates of waist circumference claim that this metric provides a more precise assessment of body fat, but a Consensus Panel in 2006 concluded that the added data would not change the recommended treatment for 99.9 percent of men and 98 percent of women. This comes as no surprise since waist circumference and BMI have a correlation of between 0.80 and 0.95. When studies discover deviations between measures of obesity, they typically occur among lighter people, or in special populations such as professional athletes; neither scenario should concern doctors who are diagnosing obesity in the general population.

When Drs. Shah and Braverman looked at the relationship between BMI and DXA, they also found a very strong correlation. In fact, substantively all of their BMI-obese patients were also DXA-obese. (See Figure 2-2.) The two metrics become practically identical, just by moving the BMI cut-point from 30 to 25.

c. The New Metric, Even if Accurately Measured, Has Dubious Value

Until a new metric generates a body of data, we cannot test its usefulness. Lots of novel measures hold promise only on paper.

The goal of measuring obesity is to combat obesity-related diseases but Drs. Shah and Braverman themselves pointed out: "Although DXA is a direct measurement of fat

---

FIGURE 2-2 Region of Disagreement between BMI and DXA: A BMI-obese patient is almost surely DXA-obese; the gray cells represent disagreements; DXA classifies 10 percent more patients as obese. [*Source:* Adapted from Table 2, Shah and Braverman, p. 4.]

All Patients	DXA		
	Not Obese	Obese	BMI Totals
BMI Not Obese	35 %	39 %	74%
BMI Obese	1 %	25 %	26%
DXA Totals	36 %	64 %	100%

Female Patients	DXA		
	Not Obese	Obese	BMI Totals
BMI Not Obese	26 %	48 %	74%
BMI Obese	0 %	26 %	26%
DXA Totals	26 %	74 %	100%

and a better measure of adiposity than BMI, it is not a disease correlate." Contrast this with decades of research that have linked BMI to Type 2 diabetes, cardiovascular disease, certain cancers, and other ailments. So, while BMI may be an imperfect predictor of body fat, Quetelet's discovery is, after almost 200 years, still the better indicator of disease than DXA.

In medicine, better outcomes can arise from more diagnoses rather than improved care. The DXA metric classifies a lot more people as obese. The marginal person being considered obese is not as fat as the average obese person, and so faces lower health risks than average. With no change to treatment methods, outcomes improve just because the expanded treated population is a priori healthier!

d. Any Measurement System Becomes More Complicated and More Costly Over Time

The urge to tinker with a formula is a hunger that keeps coming back. Tinkering almost always leads to more complexity. The more complicated the metric, the harder it is for users to learn how to affect the metric, and the less likely it is to improve it.

Getting on a scale is free and easy. Anyone who owns a basic calculator can compute the BMI formula. With pen and paper, one can chart progress. DXA scans cost hundreds of dollars, and require a visit to a clinic that has invested in expensive equipment, where only qualified doctors can read the result. To monitor changes, the patient must submit to periodic scans, which are repeated bouts of radiation. Nevertheless, spending money imparts a sense of accomplishment. More expensive wine is deemed better.

e. The New Metric Rings Out the Old

The more complex new metric requires new data. Usually, it is impossible to restate past data. As a result, all history must be whitewashed and measurement starts from scratch. How convenient! In Chapter 6, we learn that in 1994, President

Bill Clinton authorized significant changes to the survey that determines the nation's unemployment rate. New questions were added to study the habits of people looking for jobs. Since those items did not exist in the prior survey, there is a set of unemployment metrics for which we cannot make historical comparisons.

Body Mass Index (BMI) is the global standard of measuring obesity. Since the 1970s, health organizations around the world have collected data, and medical researchers have documented the association of BMI with various health outcomes. We can conveniently compare countries in a given year, or track particular cohorts within a country over time. If BMI is now abandoned in favor of DXA, which is much more complicated and costly to obtain, we are forced to erase the historical record.

4. What Is the Problem?

Shrouded in the fog of war, we are losing sight of the problem we're trying to solve. Obesity is not the adversary; rather, it is early death caused by obesity-related diseases such as diabetes and stroke. This distinction is crucial. We can win the battle against obesity and still lose the war on mortality.

In 2002, a research team led by Dr. Tobias Kurth from the Harvard School of Public Health analyzed data from the U.S. Physicians' Health Study (PHS), concluding that BMI is linked to the risk of stroke:

> Compared with participants with BMIs less than 23, those with BMIs of 30 or greater had an adjusted relative risk of 2.00 [. . . 1.48–2.71] for total stroke . . . each unit increase of BMI was associated with a significant 6% increase in the adjusted relative risks.

This language is standard in the medical literature. In plain English, the researchers are saying:

In the pool of people we studied, stroke hit obese men at twice the rate as men with BMIs under 23. The risk increments by 6 percent for each 7-pound gain in weight (for a man of average height).

In this study and many others, we're told fatter people have more diseases, and that obesity shortens lives. The statistical meaning of such statements, however, is often misconstrued.

Editors of scientific journals insist on the insertion of the word *significant*, and casual readers take this as a hint that the word provides pomp. It typically confers a statistical significance at the 5 percent level. Statistics is about variations. In this case, *statistical* concerns the variability between the men included in the PHS sample and those not included. A *significant result* is one we can apply to anyone outside the original study so long as they share a profile with the participants; in this case, they are highly educated, white men with generally normal weight, and ready access to medical care. Statistical significance does not express whether the result is important; it only tells us the result is general.

We have to look elsewhere to learn the true value of such a study. How scary is the 6 percent increase in risk? The reference here is the group of men aged 40 to 84 with BMI below 23. In any year, 0.23 percent of these men, or 30,000, are expected to suffer a stroke. Relative to this baseline risk, obese men face a 6 percent risk premium, which equates to 14 more strokes per 100,000 people. So if a study involves 10,000 patients per treatment group, we would expect one or two extra strokes. Any reasonable study would have to include hundreds of thousands of people in both BMI <23 and BMI >30 groups, or else just a handful of cases could be what separates the two groups. What journals accept as "significant" may be a tiny number.

Moreover, the medical community wields an extremely blunt instrument to perform microscopic surgery. In order to avert strokes in 28,000 men, doctors are targeting 23 million.

Such imprecision presents a problem as most treatments have side effects. Take the example of Acomplia, a diet pill marketed in Europe but withdrawn in 2009. Clinical studies found that 15 percent of the patients developed nausea; almost half the participants experienced heightened anxiety and depression, relative to 28 percent given placebo; the suicide rate of pill takers was double those using placebo. The number of patients getting adverse results is clearly much larger than the few who can benefit from Acomplia. That's why its license was rescinded in Europe, and approval stalled in the United States. Drs. Shah and Braverman's proposal to use DXA would qualify more patients for treatment, and make it even harder for the statistics to work.

5. What Is the Real Problem?

Since considerable solid research links high Body Mass Index to premature death, public-health officials naturally view lowering BMI as sound policy. Anti-obesity initiatives, like "Healthy People 2010," always revolve around BMI goals. In an interview with the *Los Angeles Times*, Drs. Shah and Braverman blame the "baloney mass index" for the winless streak:

> Efforts to get patients to shed extra pounds have produced weight loss in the short term but fatter patients in the long run as weight is regained. Medical interventions would be more successful if, instead of focusing on weight, they encouraged patients to shift their body composition toward lean muscle mass by recommending more exercise, more sleep and more healthful [*sic*] eating.

However, the regimen of more exercise, more sleep, and healthy eating have been attempted religiously, including by the likes of Darrell Phillipson, but to no avail. The doctors are correct that focusing on weight is unwise; what they don't realize is that focusing on body composition is also misguided, in exactly the same way.

To understand why, take a step back and look at the nature of medical evidence. How do researchers decide if factor X causes disease Y? The Physicians' Health Study (PHS) is typical. In 1982, about 22,000 male physicians between 40 and 84 years old enrolled in a study to determine if aspirin or beta-carotene could prevent cardiovascular disease or cancer. Which treatment a participant received was determined by a random draw. In 1992, the researchers reported that beta-carotene supplements do not prevent cancer.

The data from PHS were later appropriated to study a different topic: the relationship between BMI and stroke. Body Mass Index is obtained from the heights and weights reported at the start of the study by participants who were asked to fill out a questionnaire on their medical history, lifestyle, and personal particulars. Each year until 1995, the participants in the study disclosed whether they had been diagnosed with new conditions, including stroke. The researchers culled other information from these questionnaires, such as smoking habits, consumption of alcohol, age, and hypertension, as these are known to affect the risk of stroke in men.

The random assignment of participants to treatment ensured that those taking aspirin or beta-carotene were comparable to those given placebo in every possible way before the experiment began. If, after the experiment, one group looked different from the other, we knew that aspirin or beta-carotene caused it since the supplement was the only factor that differed. Now, this was the intent of the original research design.

Subsequent studies, such as the analysis of BMI and strokes, had nothing to do with supplements, and thus those results aren't protected by the research plan. They are known as *observational studies*, and must be interpreted narrowly. While obese men in the study face a higher risk of having a stroke, observational data cannot explain why this is so.

This point that *correlation does not imply causation* is too elementary to be missed by editors of medical journals. So, in summaries of results, one encounters perfunctory disclaimers such as:

> "We . . . analyze the association [not the cause–effect] between BMI and stroke," and "the mechanism by which BMI affects stroke risk independent of established risk factors, such as hypertension and diabetes is not fully understood."

Disclaimers are invented to keep American lawyers happy. Here, they are used to hold statisticians at bay and keep them from acting like spoilsports.

Hold your guns, they say; we know we shouldn't muddle up correlation with causation. Then invariably, they do it anyway. In the BMI–stroke study, after conceding that they didn't know the "mechanism" that causes stroke, the researchers snuck in this passage:

> These results suggest that individuals and their physicians should consider increased risk of stroke another hazard of obesity. Prevention of obesity should help prevent risk of stroke in men.

In that final sentence, the last one in the report, the cause–effect they previously admitted was "not fully understood" has raised its status to a medical prescription. Weight reduction would reduce the likelihood of stroke only if we accept that obesity causes strokes! When amplified in the media, this type of medical advice strikes casual readers as supported by scientific evidence. In fact, the researchers have evidence of correlation only, and then feed us "causation creep."

When Dr. Braverman complains about weight being the wrong target, he's really stating that weight does not directly cause Type 2 diabetes, heart disease, and so on. He's reacting against public-health officials who, inadvertently or not, succumb to causation creep. Weight or BMI is merely a marker of potential ill health. The catch is that DXA, the purportedly superior alternative, is a different marker. Neither is it a direct cause of diabetes, heart disease, and so on. The bridge from cause to effect is built on theory. It is imperative for us to recognize which part of the analysis rests upon data, and which part is strictly theory.

6. Locking in the Losses

We can't solve a problem if we are not clear what is causing it. This is the biggest obstacle in the war against obesity. Several theories blaming abdominal fat deposits are unproven. Medical science has unearthed a collection of other culpable factors, including:

- Genes
- Physiological status
- Environmental effects
- Social influence
- Individual behavior

Existing treatment strategies all suffer from the Achilles' heel of relapse. The weight regain is taken for granted. In order to get a diet pill approved, a drug developer only needs to demonstrate its effect within a 12-month window.

But there is one treatment that doesn't have the Achilles' heel. It is not for the fainthearted, though. That treatment is bariatric surgery, the most common form of which is *gastric bypass surgery*. Surgery produces dramatic, rapid weight loss beyond anyone's imagination—over 60 pounds within a few months is quite common, and many patients have successfully stayed at their new weight. In the Swedish Obese Subjects study, average loss was 61 pounds two years after operation, compared to just one pound in the control group. Better still, patients also experience attenuation in obesity-related conditions such as diabetes, hypertension, and sleep apnea. You should now skip forward in this text if you are skittish about this subject.

A white glow shines on the patient's round, gas-pumped tummy. Five cylindrical ports, not more than a half-inch in diameter, make a circle around it. Instruments travel the length of the tubes into the abdomen. The liver is folded away and strapped to the side. Cushions of fat are pushed away to expose the targeted organ, the stomach. A gadget that combines tweezers, scissors, and staple gun arrives. It is the star of the show. In one swift motion, the stomach wall is snipped not far from where the esophagus enters. When the arms of the gadget open up, a neat line of little staples hems the edge of the cut, like threads on the legs of jeans. The gadget makes its way across the stomach, cutting and sewing, creating a chunk about the size of an egg from the top. This is the new, shrunken stomach.

The small intestine is tugged out from the flabs. It is severed at 18 inches. Now, the lower portion of the stomach, including a length of intestine is sealed and decommissioned. The remaining segment of the intestine is sewn to the new stomach, forming the gastric bypass: food will get clogged up in the downsized organ, and what comes out moves through a shorter intestine, retarding absorption.

Some stray staples are swept away. With needle and threads, the orientation of various organs is stabilized. Air is pumped into the digestive tract while the abdominal cavity is flooded. Luckily, no bubbles are formed, confirming that all holes have been sealed. The intestine and liver are rearranged to their canonical posture. With such a major surgery, the patient sometimes stays in the hospital for a few days.

On returning home, the patient takes a large dose of painkillers and waits for the wounds to heal. The body slowly recovers from the shock of unplanned and extreme bowel movements. The new stomach stumbles to figure out what it is doing.

The patient takes a liquid diet. At first, it takes most of the hour to drink two ounces of milk. Some days, even that amount is unwanted, and thrown out by the uncooperative organ. The patient could go days without eating. For the first time ever, the patient cannot feel hunger, and must mindfully eat. The lungs too must adapt to the new body structure. The patient may run out of breath, requiring strengthening exercises.

The next few weeks are filled with moments of panic for the patient. It's the fear of being one of the 4,000 deaths. (About 200,000 weight-loss surgeries are performed each year in the United States, with a mortality rate of 1 to 2 percent.) The most common cause of death is stomach leakage. The patient notices a bloodied shirt, a pool of red seeping out of the surgical wound. Is the new stomach leaking? Or is the body just cleansing the insides? Every so often, the abdominal pain becomes unbearable despite the regimen of drugs.

After hours of waiting in an emergency room, and a battery of tests, the patient goes home with more painkillers, but no explanation. That is really good news. One in five patients will end up in hospital the year after surgery for more surgery. For instance, hernias may develop when the intestine backs into new crannies, necessitating operation. Gallstones are another complication.

When solid foods return to the diet, the relearning process continues. The patient discovers the agony of ingesting even one morsel of food over the stomach's limit. Certain foods are avoided, as they cause adverse reactions. Because food now takes a shorter path through the bowel, insufficient minerals are absorbed. The patient takes vitamins and supplements as a routine.

What makes these people go through so much pain and suffering? They live for the one moment each week when they get on the scale. As their bodies transform, the weight is disappearing, often at an amazing speed, five or ten pounds a week. Many patients eventually achieve a reduction of 60 percent or more. Those who can afford it elect to surgically remove the excess skin.

In his career, Darrell Phillipson, the King County judge, made decisions for a lot of people. The district court carries a yearly load of thousands of cases. When the 63-year-old retired in 2011, after 27 years of service, he made a life-changing decision for himself. He has struggled with his weight for over 40 years, attempted all kinds of treatments and diets that never produced sustainable results, and he was determined to end this suffering. He decided to go under the knife. This is an expensive choice, costing $20,000 initially, and potentially much more to address complications. It took two years to sort out the insurance coverage. For someone in his sixties, such a major surgery carries a tangible risk of death. During the first six months after the gastric bypass surgery, Phillipson frequented the hospital. Between stemming a life-threatening gastric leakage, taking out kidney stones, clearing a blockage, and so on, he returned to the operating table six or seven times.

Before the surgery, Phillipson weighed 425 pounds, and had a BMI of 63. By July 2012, he had shed 180 pounds, cutting his BMI to 36. He said his weight is still falling.

MARKETING DATA

How Can Sellouts Ruin a Business?

On May 4, 2011, Felix Salmon, the finance blogger for Reuters.com, put up a post with the intriguing title of "Grouponomics," which probably marked the premiere of this word. His first sentence was:

> Eighteen months ago, Groupon didn't exist. Today, it has over 70 million users in 500-odd different markets, is making more than a billion dollars a year, has dozens if not hundreds of copycat rivals, and is said to be worth as much as $25 billion.

Six months later, on November 4th, Groupon raised over $700 million in its initial public offering (IPO), implying an enterprise value of about $16 billion. While not as stupefying as $25 billion, $16 billion is still a handsome number, even by the yardstick of U.S.–style mega-corporatism. On the day of its debut, the value of Groupon promptly surpassed those of household brands as varied as The Campbell Soup Company, Aetna, Limited Brands (includes Victoria's Secret, Bath & Body Works, etc.), Northrop Grumman, and Intuit.

Little Groupon does only one thing. It sends e-mails to people selling "deals," deep discounts usually of 50 percent or higher on sundry goods and services. A typical deal is "[Pay] $15 for $30 worth of contemporary fare at Giorgio's of Gramercy." (See Figure 3-1.) Under this arrangement, Giorgio's cooks and serves the meal; Groupon offers the diner some free money to spend while it takes a share of the restaurant's receipts.

Groupon's one-line business is simple to describe but hard to grasp. The stock market crowned the Chicago-based company as the latest and hottest technology start-up, but it doesn't feel like one—unless we are watching when the rising red ink will breach the flood stage. Groupon accumulated over half a billion of losses in its first 18 months of existence.

FIGURE 3-1 The Groupon Deal Offered by Giorgio's of Gramercy in January 2011

Browsing the financial statements, we discover that Groupon looks nothing like a Campbell Soup or an Aetna, as it does not make goods or provide services (except e-mailing). Its shopping list consists of three items:

- Advertisements aimed at consumers
- Salespeople calling on merchants
- Copycat rivals for quick expansion

For the greater part of 2011, Groupon's bankers were busy talking up the popular daily-deals business in anticipation of its IPO. The press was evenly split between plugs and snubs. Known for his low tolerance of bullshit, Felix Salmon was wondering why Groupon deserved such a generous valuation. More than likely, this question crossed your mind too. A Google search for "groupon ponzi scheme" yields 190,000 matches.

Loyal readers—me included—expected "Grouponomics" to contain the usual incisive analysis, wit, and no-nonsense sensibility that has made Salmon one of the most lucid observers of the business world circa the Great Recession. We are accustomed to the hard-to-please Felix, who penned such pieces as: "How Rajat Gupta Corrupted McKinsey," "Adventures with CNBC's Anchors' Statistics," and "The Hermetic and Arrogant *New York Times*"; all were published within weeks of his exegesis of Groupon. But the British-born journalist blindsided us. Salmon wrote thousands of words to convince himself that "Grouponomics" was viable. To argue his case, he evoked a string of extreme descriptions, such as:

- "more or less *unprecedented* in the world of marketing and advertising,"
- "*orders of magnitude better* at targeting than anything which came before it" (to be precise, one order of magnitude is 10 times better)
- "*uncommonly large number of ways* in which participating in a Groupon deal can benefit a . . . merchant"
- "a positive reputation which can *spread like wildfire*

over Facebook and other social networks" [All italics are mine.]

Beneath that oddly cheerful veneer, Salmon constructed his argument:

> Groupon has created a two-sided marketplace, connecting consumers and local merchants; consumers are all too eager to save money, and merchants, to pick up new customers; so long as both sides fulfill their objectives, Groupon can collect the brokerage fee.

To Salmon, Groupon is much more than a digitally transmitted coupon, because of personalization: "the more people Groupon signs up, the more targeted its deals can be." Groupon uses an algorithm to enhance the relevance of deals shown to its users. Perfecting this capability will be crucial to the company's long-term prospects. Salmon reiterated this point in a post called "Whither Groupon?" four months after "Grouponomics," when other observers started to worry about a reported slowdown in Groupon's rate of revenues.

Perhaps Groupon sends you e-mails. Perhaps you are suspicious if the free money comes with a trap. Perhaps you've redeemed several deals. Perhaps you considered buying GRPN stock at IPO. Perhaps you've heard stories of merchants who almost met their ruin after running Groupon promotions. Perhaps you're tempted to short the stock now. How does NUMBERSENSE help you take in the Groupon juggernaut?

1. The Fine Line between Profit and Loss

The masterstroke in Felix Salmon's "Grouponomics," his journey to the heart of Groupon's business model, was spending half of the blog post discussing the merchant's experience, primarily that of Giorgio's, a restaurant in his Manhattan neighborhood. He nails the key issue: It matters little how much consumers love Groupon deals, and love them they indeed do, but an exodus of merchants could spell doom to Groupon's marketplace.

Even experienced columnists are blind to the merchant angle: Check the review by David Pogue, the celebrated *New York Times* technology reviewer and another fanboy of the $16-billion start-up, in which he gushed superlatives, describing a rush of feelings including "giddiness," "thrill," "exclusivity," and "serendipity." The one time he mentioned merchants was in a sweeping declaration that they "pick up new customers overnight without doing a lick of marketing of their own." This last statement celebrates getting something for nothing. If a restaurant owner sells out his dining room through free advertising, what's not to like?

Most people have a general notion about how merchants make money from offering deals through Groupon. The basic math, which serves as our starting point, was presented to potential investors during Groupon's IPO roadshow as a case study.

Seviche restaurant, in Louisville, Kentucky, sold about 800 coupons with the deal "$25 for $60 worth of Latin cuisine and drinks." The average Groupon user typically spends $100 at the restaurant. He or she pays with the $60 coupon and $40 out of pocket plus any taxes and tips. Seviche keeps the entire $40, and later collects $12.50 from Groupon, which represents a 50-50 split of the $25 the diner previously paid to purchase the coupon. In sum, Seviche makes $52.50 from the meal. After subtracting $33 it costs to serve the food, Seviche's gross profit is $19.50 per table. If all 800 coupons are redeemed, the total gross profit amounts to about $15,000. This amount, Groupon reminds the merchant, ignores future visits by the same diners.

One glance at this basic math, and it's not a stretch to conclude, as Pogue did, that merchants like Seviche can "pick up new customers overnight without doing a lick of marketing." Magical Groupon delivers customers for free! But think twice about those rosy numbers.

Take a look at Figure 3-2, which shows a different view of the math. For a regular diner, a check of $100 results in a

FIGURE 3-2 The Case of the Missing Revenues: Seviche takes in $19.50 from each Groupon user, compared to $67.00 from a regular diner. Where did the money go?

	Regular Diner	Groupon User	
Check size	$100.00	$100.00	
Cost of meal	−$33.00	−$33.00	
	—	−$47.50	← *Where did this go?*
Profit	$67.00	$19.50	

gross profit of $67. For the Groupon diner, a check of $100 yields a profit of only $19.50. An amount of $47.50 has gone missing. Where's the money? Groupon claims $12.50 while the other $35 flows to the customer. It's a $25 for $60 deal after all. So Seviche could have earned $67 but actually takes in less than one-third of that amount. What Groupon touts as a winner could really be a dud.

The $19.50 may have seemed like a decent profit but next to $67, the what-could-have-been, it looks meager. The what-could-have-been is what statisticians call a *counterfactual*, and it's one of the fundamental constructs in statistics. If the same customer had dined at Seviche without presenting the coupon, then Seviche would have earned $67 that night. In reality, the restaurant made only $19.50.

Since the official analysis uses the actual amount received ($19.50) as the "ground truth," you might feel it can't be wrong. But you'd change your mind if you're aware that the simple math conceals a bold assumption: Every coupon user is a *deal seeker* who dines at Seviche solely because of Groupon's discounting. If 50 diners redeemed coupons on a given night, we made believe that those 50 tables would have remained unoccupied if Groupon didn't exist. That's implausible.

Did Seviche win or lose? In the official story, the restaurant is a clear victor. In my version, the Groupon promotion has the effect of splitting Seviche's potential profit in three, one part to the diners, one part to Groupon, and the third retained by the restaurant. The truth lies in the middle. Some Groupon buyers are *newbies*, who have never visited Seviche, while others are *free riders*, who dine there regularly. It is the *newbies-to-free-riders ratio* that determines the level of profitability, thus satisfaction, of merchants.

Each free rider inflicts a loss of $47.50 that can be covered by incremental profits from newbies, at $19.50 per table. The economics balance if the promotion attracts at least 2.5 new customers for each regular customer. Said differently, 70 percent of redemptions must come from first-time customers in order to break even. (Ouch!)

But surely, you protest, 800 Groupon customers spent $80,000 at Seviche, bringing the restaurant $15,000 of profit, in addition to filling up empty tables. Why would I fret over an imaginary loss? Let's assume half of those diners were newbies and the other half were free riders. If Seviche didn't use Groupon, the restaurant's gross profit would have come from 400 diners at $67 each, totaling $26,800. (The other 400 tables would have contributed nada.) So Seviche left $11,800 on the table!

Such an experience demoralizes the unsuspecting business owner. At Posies Bakery & Cafe in North Portland, Oregon, Jessie Burke ran a deal offering $6 for $13 worth of food. Three months later, her business appeared healthier, and yet she lost so much money she had to inject $8,000 to pay rent and wages. "It was sickening," she described the unexpected loss, "especially after our sales had been rising."

2. What Could Have Been

The industry most commonly associated with Big Data is online marketing. E-commerce websites generate amazing volumes of data 24 hours a day, seven days a week, as site owners watch over every finger tap and every mouse slide. The day of the anonymous customer has passed, as credit and debit cards, and electronic payment systems must verify names and addresses. Big Data is why online marketing and advertising are supposedly more measurable and seriously more accountable than traditional marketing and advertising. Experts in this emerging area frequently flunk the counterfactual test. Let's look at two examples.

a. Dell Computer and the Benefit of Twitter

The hip business magazine *Fast Company* sneered: "All you doubting Thomases can shut up now: Lifecasting/social net Twitter really does work as a marketing tool, as confirmed by PC retail leviathan Dell." Dell, an old guard of the U.S. technology industry, just offered an endorsement to the sector's newest starlet, Twitter, in 2009. Twitter is a flourishing Web service that fascinates the early adopters as much as it puzzles bystanders. Superficially, Twitter puts people's text messages online, now relabeled "tweets." A hitherto private medium is turned public. Anyone can "follow" anyone's stream of tweets. Particularly witty messages are "re-tweeted" to one's roster of followers, similar to forwarding e-mails with especially raucous jokes among friends. When users enter their Twitter accounts, they eavesdrop on a super-feed of all the messages dispatched by people they follow. Logging on to Twitter produces the sensation of wandering inside a packed restaurant on a Friday evening and hearing all of the conversations at once.

As a direct-to-consumer retailer, Dell can't wait to join those conversations. Within two years of launching @Dell-Outlet on Twitter, the vendor sold $6.5 million worth of computers, accessories, and software. The 1.4 million followers of Dell's Twitter presence learn about special offers from 35 streams in 12 countries. *Fast Company* offered a standard analysis of the return on investment (ROI):

If Dell pays each of its 100 Twitter writers an average of $65,000 per year (benefits and overhead costs included), and she spends 20 percent of the day crafting 140-character short messages, then the computer vendor invests $1.3 million annually in its Twitter marketing program, implying an attractive return of 150 percent (half of $6.5 minus $1.3 divided by $1.3). Each $100 expense produces $150 incremental revenue.

A data analyst at Dell can link every dollar of the $6.5 million to a sequence of actions; he can locate to the hundredth of a second when a customer authorized the credit-card transaction, when a customer accepted the terms of sales, when a customer placed the merchandise in the virtual shopping cart, when a customer arrived at the e-commerce website, and most tellingly, when a customer clicked on the Twitter message. Isn't every dollar honestly earned if it can be traced to a click on a tweet created by one of the 100 writers? Online marketers worship the clickstream as the Holy Grail. How much more proof of success can one demand?

Whenever shown ROI numbers, we should ask about the counterfactual. If Dell's marketers rebuffed Twitter, would the entire $6.5 million worth of sales implode? We can't observe that alternative reality but a good guess is no. No one becomes an @DellOutlet follower by accident; you actively subscribe to Dell's Twitter stream. So followers are shoppers on the hunt for a new computer, including fans of Dell's well-established reputation for high quality and fair pricing. They are currently seeking bargains, and understand the short shelf life of tweets. Were Dell to terminate its Twitter presence, most of its followers would have purchased Dell computers anyway, and that's because the alternative to Twitter isn't silence—Dell reaches out to these customers also through catalogs, e-mails, retail stores, product placements, TV commercials, and other channels. Thus, the reported $6.5 million in sales, and 150 percent ROI are vastly exaggerated.

How many of those people would punish Dell for not being on Twitter?

Statisticians set a high bar when they assign a cause to an effect. The most popular standard is the counterfactual perspective, doggedly championed by Don Rubin of Harvard since the 1970s. The impact of Dell's Twitter program is only a fraction of the sales tied to the clickstream. We need the imagined world to help interpret the real world. The imagined world is the counterfactual one in which Dell did not tweet. Dell's marketers build numerous paths to the sale, so that blocking one still leaves other roads open. The customer could, for instance, dial the customer-service line, or go directly to Dell's website to initiate a transaction. The Twitter writers earn their salaries by hooking *incremental shoppers* whom Dell could not reach through these other paths.

Counterfactual thinking makes it clear that the clickstream is not causal. The sequence of clicks identifies the path by which the sale occurs, but it's a mistake to confuse the *how* with the *why*.

b. IDC and Cost of Software Piracy

IDC (International Data Corporation), a leading market research firm, could have avoided embarrassment if it had looked at what could have been. The Business Software Alliance (BSA), a lobbying group for the software industry, pays IDC to produce an annual report on global software piracy. In this report, analysts estimated the monetary loss to the software industry due to piracy. Using various surveys, the research firm determined the volume of new software that was pirated, and multiplied that number by the average retail value of software. The firm insisted on calling the result "piracy losses" until 2009 when it switched the terminology to "commercial value of unlicensed software."

The relabeling reveals a conversion to the counterfactual view. Critics of the first five reports charge that a significant proportion of the apparently real demand for software would not have materialized in an imagined world. A lot of alleged users of pirated software, especially those living in poorer parts of the world, would have simply done without the software if piracy were somehow eradicated. Thus, not every dollar worth of unlicensed software constitutes a direct loss to the industry. Free leads to overconsumption: This is why the owner of a restaurant serving Asian-style buffets I once visited in London posted a cheeky sign, warning "One pound for each noodle left in your bowl."

To properly evaluate the impact of piracy, one must imagine what the world would be like if software could not be pirated. Some degree of guessing would be required, but ignoring this imagined world leads to incorrect conclusions for sure. When in doubt, ask what could have been.

3. The Importance of Typecasting

Jessie Burke, owner of Posies Bakery & Cafe, was shaken by the unforeseen losses arising from Groupon deals when one of her best customers showed up—one day late with her expired coupon eager to have it redeemed. She reluctantly turned her away. Of course, the customer was offended. (They did make up after Burke opened up about her ill-fated dalliance with Groupon, later shared on the restaurant's blog.)

Despite some unpleasant encounters that ruined her overall experience, Burke recognized that the bargain-basement promotion introduced "many, many wonderful new customers" to her cafe. She discovered there is no such thing as the "average" Groupon customer.

At EaT: An Oyster Bar, also in Portland, Oregon—one of Groupon's most successful local markets—there were 1,500 takers of coupons offering $25 of seafood for $12. The response overwhelmed the owners of the three-month-old restaurant who detailed the scene:

> Swarms of first-time customers (most of whom never came back again) crowded out, undercut and alienated our regulars who were paying full price. Servers got stiffed on tips.

The loyal customers probably felt anger mixed with disgust, like when you are on a flight, and the friendly passenger across the aisle discloses that she paid half of your fare.

Both merchants instinctively recognize two types of coupon users, which can be labeled the *newbie* and the *free rider*. Marketers call such typecasting *customer segmentation*. One difference between the two segments is easy enough to spot: The free rider would have made a purchase anyway while the newbie appears only due to Groupon. The earlier analysis shows that each first-time customer provides incremental revenues, albeit at a brutally reduced margin, and that every free rider imposes a loss. The two segments diverge in other more subtle ways as well. Seasoned marketers take note of these factors when designing promotional tactics.

The free rider is less likely to squeeze the merchant by leaving miserly tips, or by spending not a cent more than the face value of the coupon. As a regular customer, he or she is cordial with some of the staff, and knows what to spend money on. The anonymity of the first-time visit encourages people to behave poorly, basing tips on the bill excluding the deep discount, re-using coupons, obtaining multiple coupons, and so on. They will bend the rules if they don't plan on returning.

Ironically, it is the free rider who loves the deals more. As Felix Salmon mused, "If you're already a regular [customer] somewhere, of course, then buying its Groupon is a no-brainer." The *New York Times* reviewer David Pogue confessed to free riding, as he blissfully described buying $10 for $20 worth of Italian food at a neighborhood restaurant, $15 for $30 worth of dry cleaning, and $10 for $20 at Barnes & Noble "since that's all stuff I'd buy anyway." Because

Groupons require prepayment, newbies may think twice before acting on their impulse, furthering the "adverse selection" of free riders.

The average newbie has a much lower chance of revisiting the shop. The free rider, by definition, has been a satisfied customer all along, and he or she is more willing to pay in full on return trips. Especially for services like yoga classes or salons, the newbie would face sticker shock, when and if he or she returns. Hannah Jackson-Matombe, owner of Spotless Organic in London, told the BBC: "We've had very good feedback from [Groupon] customers on the whole. But if you paid £20 for my oven cleaning service that would normally cost £99, you wouldn't do it [at full price]—I wouldn't do it!"

The two concepts—the counterfactual and customer segmentation—together lead us to the following picture of merchant "Grouponomics" (Figure 3-3):

FIGURE 3-3 Merchant Grouponomics: Net revenues are gross revenues net of the Groupon discount. A third group of customers, regular customers who don't use Groupons even when available, contributes the same revenues under either scenario, and thus does not figure in this analysis.

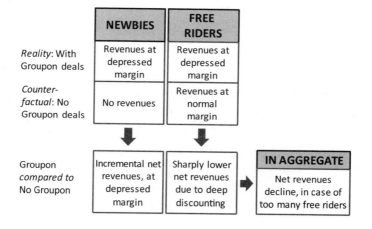

	NEWBIES	FREE RIDERS	
Reality: With Groupon deals	Revenues at depressed margin	Revenues at depressed margin	
Counter-factual: No Groupon deals	No revenues	Revenues at normal margin	
Groupon *compared to* No Groupon	Incremental net revenues, at depressed margin	Sharply lower net revenues due to deep discounting	**IN AGGREGATE** Net revenues decline, in case of too many free riders

Contrast this with the way-too-simple official analysis shown in Figure 3-4:

FIGURE 3-4 The Official Analysis Is Too Simple: The official analysis, accepted by the mass media, fails to capture the experience of merchants who lose money even as they improved sales.

	IN AGGREGATE
Reality: With Groupon deals	New customers produce net revenues that are profitable after costs

---

If you are the merchant, you want to clog cyberspace with coupons to reach as many new customers as possible, and you wish that your loyal customers will stay unaware. These two goals don't sit well together. You want to hire a rainmaker, but you want the manufactured drops to fall only in specific locales. *Targeting technology* is supposed to resolve the conflict. Think of targeting as a mechanism for sorting. If Groupon can invent algorithms to target the desired set of buyers, the deals would be a net positive for retailers. We'll pick up on this topic in Chapter 4.

4. Toying with the Model

The actual experience of a Groupon retailer is not uniform. That much we know based on a stream of reviews that have appeared in press. Some merchants vow never again to use Groupon, while others feel Groupon has delivered everything they wish for. In response to critics, Groupon's defiantly irreverent founder, Andrew Mason, circulated a pep-rally-style memorandum amongst his troops in August 2011, boasting that "the negativity leaves us well-positioned to exceed expectations with an IPO baby that, having seen the ultrasound, I can promise you is not one of those uglies." The dissenters, Mason suggested, defeated his earnest effort to elucidate:

> I tried my best to explain [the Groupon story] simply,
> but it's not lost on me that if you actually understood
> this, you probably had to read it three times. It's not
> easy stuff. It's much easier to assume that we're goons.
> So people can be forgiven for being suspicious.

What does our "toy model" say about the controversy over the value of Groupon for its merchants? The profitability of a deal hinges on the balance between two types of customers, as measured by the newbie-to-free-rider ratio. Stores with the right characteristics make money from these promotions but Groupon is hardly a godsend to all. The toy model gives us useful clues about who can take the greatest advantage.

Any store with few regular customers, such as a new business, has good odds to be a winner. When only a limited amount of revenues can be lost to free riders, most coupon users will be newbies. In March 2011, Jason Waddleton, who runs The Haven, a Scottish pub in Boston, used Groupon specifically to promote a just-launched lunch-and-brunch service, and it was a smash hit. The 60-seater restaurant sold 1,300 half-off coupons. Waddleton greeted the onslaught of diners: "Welcome to our busiest-ever brunch!"

Any retailer who can count on future visits stands a bigger chance of winning. The toy model can be augmented to account for future revenues, known as *lifetime value*. Because the average newbie spends more money over time, fewer of them are needed to fund the giveaway for free riders. Assume, optimistically, one-third of the first-timers will return to Seviche restaurant over the next year, and if they do, they will dine twice and pay full prices. Then on top of the $19.50 earned from the first visit, Seviche can expect to receive another $44.70 (33% × 2 × $67) over the year. Now, each newbie pays for 1.4 free riders. This still means 4 out of 10 Groupon users must be newbies to make the merchant whole.

Achieving even that ratio proves to be a formidable task for some stores. When U.S. Toy Company offered deals of $5 for $20 in their Kansas City location, they noticed that 90 percent of the Groupon users were existing customers, and the store lost money on three-quarters of the redemptions. Besides, the deep discount on the first visit may have set an image of cheapness in the customer's eyes, causing sticker shock when he or she next faces the prospect of paying regular prices.

Rather than counting on return visits, the merchant can sometimes supersize the first sale. Seviche restaurant can entice diners to spend more than the face value of $60. Ordering a bottle of wine will typically be enough. At one massage parlor, customers learn that they have a choice of giving up $11 of the advertised $50 savings or adding extra services, which require overspending by at least $10. Our toy model incorporates this overspend factor. For Seviche, an average bill of $100 equates to an overage of $40 above the coupon's face value. The economics improve or worsen as the overspending grows or shrinks.

Some retailers may have trouble enticing shoppers to overspend. While one can imagine splurging at restaurants, a yoga studio that is giving away 10 classes for the price of two may have trouble selling more on the first day, particularly if the clientele comprises many deal seekers. What about stores selling things? Jonathan Freiden, the third-generation owner of U.S. Toy Company, for instance, told the *Wall Street Journal* that most of the 2,000 people who took advantage of his deal "didn't spend even our average sale. It was just sad." There is also a practical difficulty. Ms. Jackson-Matombe of Spotless Organic explained: "You're suddenly inundated [with coupon users], and you have no chance to upsell anything."

Overspend has a flip side. From the consumer's perspective, each dollar of overage is a dollar not saved. Consider Seviche restaurant's offer of $25 for $60 worth of food, which was advertised as a 58 percent discount. If you spend $100, your offer is effectively $65 for $100, more honestly depicted as 35 percent off. Smart users will soon figure this math out.

Some observers claim businesses with fat gross margins can absorb the cost of Groupons. That's not what the toy model tells us: Expanding Seviche's gross profit margin from 67 percent of revenues to 85 percent does not alter the $47.50 loss per free rider. If this doesn't sound right to you, remember that in the counterfactual, the free rider would have contributed a gross profit of $85 if he or she had shown up without the coupon. If your business is enjoying super margins even before you engage Groupon, it will still be super-profitable without Groupon. In fact, your fat margin would compress a little after Groupon takes its share, wouldn't it?

Now, coupons do become more affordable when fewer users redeem them. If deals are purchased and never used, the *paid-in value* drops directly to the bottom line of the merchant, except in those cases in which Groupon hoards the expired value. However, because Groupon demands prepayment, the rate of redemption is sky high, typically above 70 percent, and hard to control. For contrast, shoppers claim less than 1 percent of the cents-off coupons for cereals freely distributed in U.S. newspapers. In the end, it's quite sad to build a business that bets on amnesia. It's also improbable that the same customers who forget to redeem the coupons would keep buying them.

Once we have a toy model, it is a simple matter to explore various settings:

- Less overspend
- More redemptions
- Higher margins
- Fewer return visits

Data analysts start with these toy models. Then, they peer outside the lab to check how closely their worldview mirrors reflect actual experience. On finding misfits, the analysts adapt their models by adding layers of detail.

A Groupon-style promotion makes sense for merchants who can find the right balance between newbies and free riders. One thing is for sure: Groupon is not free advertising. It is free only if you ignore the free riders. It is free only if you want to donate some of your hard-earned profits to the high-profile tech company and its deal-hungry clientele.

If I were to speculate, Groupon can sustain a niche business similar to Restaurant Weeks. Two types of restaurants figure prominently in the list of participants. New entrants, with few loyal customers, have little to lose, and often offer the best value for money. Some established restaurants use the promotion to fill seats cheaply during the low season. These are the places that serve special menus offering anything but their regular fare. These are the places that can make money from those one-time meals. They aren't counting on the return business.

Will Personalizing Deals Save Groupon?

You had a rough day at work. At 6:05 p.m., you were about to leave the office, and as you were putting on your coat, your BlackBerry beeped. A note from the boss: He wanted Report #10 on his desk the next morning. You imagined the bastard two-thumbing the message from his buddy's posh Tribeca pad where a rowdy night of poker was well under way. You had also made plans—dinner with your wife at the sushi joint down the block. You called home. No, you didn't. The prospect of a senseless roasting made you stop. You need something to soften the blow. You fumbled through the *Times*. What was on show at Cinema Village tonight? Bingo. A documentary about surfers in Papua New Guinea. A pair of sisters raring to break into the male-dominated culture. You rehearsed the call: apology, dinner cancelled, let's go to see *Splinters* later, surfing, third world, women's issues, rave reviews, love you. Meanwhile, a note flashed on your computer's screen. A silly e-mail from the wine-soaked boss? No, it's Groupon. Half price for *Splinters* at Cinema Village. A perfect silver lining. You clicked on the big "Buy!" button, and wished yourself luck as you dialed home.

Marketers dream of scenarios like this one. Making the right offer to the right person at the right time is the trifecta of marketing success. The Yoda level is the state of "personalization," sometimes described as one-to-one, as if the marketer is speaking directly to you. Marketers presume people don't hate advertising or promotions—not if the message is relevant. Groupon, the daily-deals company, has been likened to a gigantic database of e-mail addresses. Supporters, like the finance blogger Felix Salmon, consider its proprietary targeting algorithms to be critical to the start-up's success. How effective really are Groupon's targeting algorithms? I set out to find out.

1. Rummaging through E-mails

Augustine Fou, a veteran digital marketer, graciously let me poke around an archive of Groupon e-mails containing 776 deals he received over a six-month period from December 1, 2010 to June 30, 2011. Up until the end of March 2011, Groupon delivered one deal a day to his inbox. Since then, the rate jumped to six daily, with a featured deal appearing once by itself and once as the lead offer on top of four other deals. The offers came from all corners, from SpeedNYC Dating to Laughing Buddha Yoga Center, from the Brooklyn Film Festival to Nicholas Toscano D.D.S., from Goodfellas Pizza on Staten Island to Nutbox, a purveyor of dried condiments with various locations around New York. Deals at restaurants were the most frequent (124 instances), followed by spas and salons (85), fitness (73), and beauty (48). Among the least common were coupons from jewelry shops (1), dating services (2), and pets-related businesses (3).

I asked Fou to rate every category of merchants by his level of interest from 1 to 5. He gave 1s to restaurants, gourmet food stores, gift shops, and men's fashion stores, indicating he opens e-mails from these businesses. Since he is married, doesn't drive in Manhattan, doesn't own any pets, isn't looking to switch doctors, and doesn't enjoy dancing, he ignores all offers in any of those categories (5 rating). The in-between ratings are "sometimes interested," "neutral," and "usually not interested."

If Groupon's targeting technology is as smart as advertised, it should know that Fou loves food, and is often in the market for clothing (usually men's staples) and gifts. It should avoid sending him deals from doctors, dance studios, pet stores, and dating sites. So what does the data say?

Figure 4-1 summarizes my finding. The only clear match between Fou's expressed interest and Groupon's offering is

FIGURE 4-1 Matching Groupons to Fou's Interests: Only 34% of the Groupons presented to Augustine Fou from December 1, 2010 to June 30, 2011 came from categories of merchants that Fou rated as "definitely" or "maybe" interested.

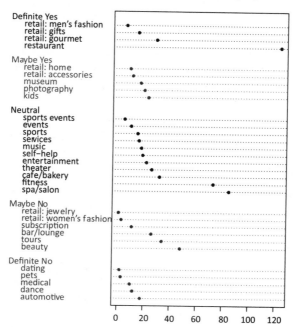

**Number of Groupons
by Merchant Category**

the restaurant category. Restaurants account for about a quarter of all Groupon deals in general; for Fou, dining offers formed 16 percent of the total. The quality of targeting does not impress. Apart from restaurants, Fou received the most offers from spas, salons, and fitness centers, all of which he views with ambivalence. Then followed beauty and touring, two categories he totally ignores. In total, only 34 percent of the deals in Fou's inbox belonged to the first-or-second-rated categories.

We expect a targeting system to improve over time. Perhaps Groupon's computers picked up clues and improved the deals over the six months. From Figure 4-2, we found scant evidence of such learning. Disappointingly, the proportion of food-related deals diminished while more weight was given to some of Fou's lesser hobbies like beauty, touring, and bars/lounges (all rated 4 out of 5).

Fou's trove of Groupon e-mails seemed to tell the story of impotence. Most of the deals that landed in his inbox were

FIGURE 4-2 Trend in Deal Types: The number of most-preferred deals, including restaurant deals, reaching Augustine Fou declined in importance over time. He received comparatively more bar/lounge deals, a category in which he had little interest.

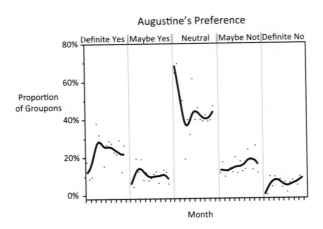

irrelevant. Should Groupon investors be concerned about this apparent failure?

2. The Joy of Failing

If you think Augustine Fou may be an outlier, you should cast aside that easy explanation. I didn't bring up Fou's experience to spite Groupon. His case is typical and predictable, if you ask statisticians who build targeting models for a living. Examine your own collection of Groupon e-mails, and you'll discover that the coupon vendor hasn't succeeded in predicting your likes and dislikes either.

Supposedly, baseball legend Ted Williams once commented that his sport "is the only field of endeavor where a man can succeed three times out of ten and be considered a good performer." A similar game of odds faces the courageous statistician who endeavors to forecast human behavior. Perhaps it is no wonder that baseball statistics so fascinate statisticians the field of study has acquired a name (sabermetrics), a Brad Pitt–headlined movie (*Moneyball*), and annual gatherings at MIT (Massachusetts Institute of Technology), which attract hundreds of devotees.

Do targeting models fail 30 percent of the time and still rate as successful? Let's find out. We can use information from Groupon's financial statements to gauge the effectiveness of its targeting technology. In the third quarter of 2011, the daily-deals company sold 33 million coupons while administering a massive database of 130 million e-mail addresses. Since each subscriber receives 30 e-mails a month, each one displaying five deals, we can compute how many Groupons were exposed to subscribers during those three months: a whopping 58 billion. Coupon sales of 33 million derived from 58 billion deals pitched equals a rate of response of 0.06 percent. That is six purchases for every 10,000 coupons presented. In reverse, that is 9,994 failures for every 10,000 attempts! This targeting business is stridently less forgiving than hitting a fastball. (The hit rate is even lower if we attribute a portion of the sales to visitors of Groupon's website rather than to e-mails.)

Whatever targeting technology deployed by Groupon circa 2011 produced the 0.06 percent rate of response. Many observers, including Reuters's finance blogger Felix Salmon, count on Groupon investing heavily to enhance this technology to justify its extraordinary valuation. (On the day it went public, Groupon was already worth more than Aetna, a health insurance giant with annual revenues of $34 billion, serving the medical needs of 35 million people.) Say Groupon develops an uber-algorithm that raises the hit rate 100 times. At 6 percent, Groupon makes six sales per 100 pitches. Marketers would be proud of achieving that level of performance even though 94 out of 100 attempts fail. Soon, we will learn that such a phenomenal rate of success comes with certain sacrifices. Presently, we investigate how targeting technologies make magic.

3. Miranda Priestly Meets Patrick Bateman

In the early 2000s, the MTV dating show *Room Raiders* infamously hanged on television the dirty linen, panties, and other gross items of teenage participants. In each episode, three youngsters vie for a date. The kids must impress the potential boyfriend or girlfriend in an unusual way: by opening up their living quarters, closets, and belongings to inspection. Having not met the contestants, the date attempts to discern their traits, habits, likes and dislikes, while rummaging through their junk. *Room Raiders* unabashedly preys on an audience hungry for guilty pleasure, but it is much more: Watching the show is like watching a targeting machine at work as it scrapes up and sorts through miscellaneous clues to form opinions about the proclivities of strangers.

Let's imagine we have a targeting machine code-named "Miranda Priestly" after the diva of fashion in the 2006 hit movie *The Devil Wears Prada*. As the consummate industry insider, Miranda has her nose on every trend and every fad over decades. Give Miranda a name, and she will recommend an outfit sure to garner praise. We lead her to the Manhattan bachelor pad owned by Patrick Bateman, the immodest banker with uncompromising fashion sense from Bret Easton Ellis's *American Psycho*. It wouldn't take a three-count for Miranda to diagnose Patrick's obsession with Armani suits, Ferragamo shoes, and Oliver Peoples eyeglasses. We give Miranda a tour of the museum holding over 1,000 pairs of shoes left behind by Imelda Marcos, the former First Lady of the Philippines, when she fled the country after her husband was ousted from power in 1986. Miranda immediately notices Imelda's love affair with shoes. Guess which section of a department store Imelda frequented? That one's easy, but what about her preference for styles, colors, and brands? This problem is trickier but still tractable; she examines a sizable sample of the shoes on exhibit. Miranda's recommendation for the late Apple CEO Steve Jobs? Black T-shirts. For Su Li-zhen, Maggie Cheung's character in *In the Mood for Love*?, Cheongsams.

For Taylor Nitiolex? Taylor who? We have never met Taylor, and we don't know his address (or perhaps hers?). Miranda has no closets to inspect. What to do now? Without clues, it seems hopeless to predict if Taylor would buy a hoodie from Hollister, or a black evening gown by Vera Wang. It is more prudent to guess among larger categories, such as men's or women's clothing, shoes, or accessories. Miranda can use the naïve strategy of picking one category at random, just like the clueless student making a wild guess on a multiple-choice test item. Allowing for the possibility that Taylor Nitiolex has no interest in fashion at all, Miranda would be picking one of seven categories, and therefore, she would guess correctly one-seventh of the time, purely by luck. But we know Miranda is the opposite of naïve. As the editor of a premier fashion magazine, she knows sales do not split up evenly among different departments. The women's fashion market is nearly double that of men's. So, Miranda predicts women's clothing for Taylor. Her instinct lifts the odds of being right above random guessing. She has thus invoked the law of averages: She regards Taylor as the "average" consumer. Miranda can do even better than this. By looking up a database of first names, she learns that three of four newborns named Taylor are girls. She should therefore think of Taylor as the "average" woman instead of the "average" customer.

Since any individual is extremely unlikely to behave like the average person, Miranda's success rate is still low. More information will aid Miranda's cause. For example, if Taylor is a 33-year-old single gal living in a rental apartment in downtown Manhattan, Miranda can treat Taylor like the *average* 33-year-old single gal living in a rental apartment in downtown Manhattan. She understands this segment of customers, and their fashion sense. Miranda's overall strategy is to place each consumer into a "look-alike" group, and treat each consumer just like an average person from that group. Any targeting technology lives and dies by its ability to uncover look-alike groups.

In the above, we ask Miranda to match all people to all things. A more realistic objective is to find targets for a single category of items. For example, to support the launch of a new line of jeans, The Gap may want to send special introductory offers to selected customers. They may approach Miranda, asking her to whittle down a long list of potential customers to a roster of the most inviting targets, those who would be most receptive to coupons. Based on data, including the relationship with the Gap brand, Miranda then rates everyone on a scale of 0 to 1, ranging from disinterest to utmost involvement. Customers with similar ratings form look-alike groups. Yes, indeed, each one of us is reduced to a number. But at least we are not mass-produced as an average person.

You see how Miranda Priestly can guide Patrick Bateman with much stronger confidence than she can guide Taylor Nitiolex. With Patrick, she draws on direct observation of his past purchases, and realizes he's a one-note guy. Taylor, by contrast, is mysterious. So is Augustine Fou to Groupon's targeting guru. He has purchased but a few coupons, and Groupon itself has had few operating years. Given the miniscule hit rate, most customers are like Fou, with little known about them, and so the deals they see tend to hit wide of the mark.

4. Where's the Target?

The most common kind of coupon received by a Groupon subscriber is the misguided one. Just like baseball players, modelers live with the knowledge of constant failing. But as statisticians, they find excitement in beating the tiny odds.

If 0.06 percent strikes you as an abysmal hit rate, imagine what can happen if Groupon switches off its targeting machinery. Had Groupon delivered coupons to a random assortment of inboxes, ignoring our likes and dislikes, they might have succeeded 3 out of 10,000 attempts. The allure of targeting is contained within that interval between 0.03 and 0.06. This 100 percent improvement is a badge of honor, even though you and I get mostly the wrong deals.

Recall the Louisville restaurant Seviche, in Chapter 3, that ran a Groupon promotion in February 2010. Groupon has 200,000 subscribers in the Louisville area, of which 6.5 percent (13,000) got Seviche's e-mail, and of those, 6.2 percent (800) made a purchase. Pretend, for the moment, that Groupon had randomly fished out the 13,000 subscribers from its vast database so that 6.2 percent is the average hit rate. Now, if Groupon had solicited all 200,000 subscribers, sales would have topped 12,300 (6.2 percent of 200,000). By restricting its promotional campaign to 13,000 e-mails, Seviche attained merely 6.5 percent of its full commercial potential (800 of 12,300).

We just emulated how statisticians evaluate targeting models. Two key points pop out of this analysis: The strategy of random selection dispatched 12,200 e-mails to people who weren't buyers (false-positive errors) even as the spray of e-mails missed 11,500 potential buyers (false-negative errors).

What if Groupon switches on its targeting intelligence? Assume the targeting strategy has triple the effectiveness of random selection so that by reaching out to 6.5 percent of its subscribers, Groupon captures 19.5 percent of purchasers (2,400 coupons sold). The modelers must work miracles to deliver this level of performance, and still success will leave 80.5 percent of potential sales on the table (12,300 minus 2,400 divided by 12,300).

To scoop up some of those missed opportunities, statisticians can expand the list of targets, say, doubling the reach to 26,000 subscribers (13 percent of the database). The restaurant could expect to move at least 1,600 coupons (13 percent of 12,300), sales that would require no special technology. Introducing smart targeting might yield another 1,230 sales (an extra 10 percent) for a total of 2,830 or 23 percent of the full potential. That the doubling of the volume of e-mails would not double sales is the *law of diminishing returns*. The first 13,000 subscribers are more prone to buying a coupon than the next 13,000, if the predictive model delivers the goods. Recall that targeting models assign each subscriber a rating, indicating his or her probability of purchase. They allow Groupon to harvest the low-hanging fruits, those customers deemed most likely to buy. Figures 4-3 through 4-5 present the details of these metrics.

The trick of sending more e-mails is a double-edged sword. While the number of false-negative errors falls, the algorithm suffers from more false alarms. In the Seviche restaurant example, false positives jump from 10,600 to 23,170. This trade-off between two types of errors also challenges designers of lie detectors, terrorist-prediction algorithms, and anti-doping tests, which I detailed in *Numbers Rule Your World*.

FIGURE 4-3 Method One of Targeting: Groupon selects 13,000 names *at random* from its massive e-mail database. The hit rate of those receiving e-mails (800/13,000 = 6.2%) is the same as those excluded (11,500/187,000).

RANDOM SELECTION OF 13,000

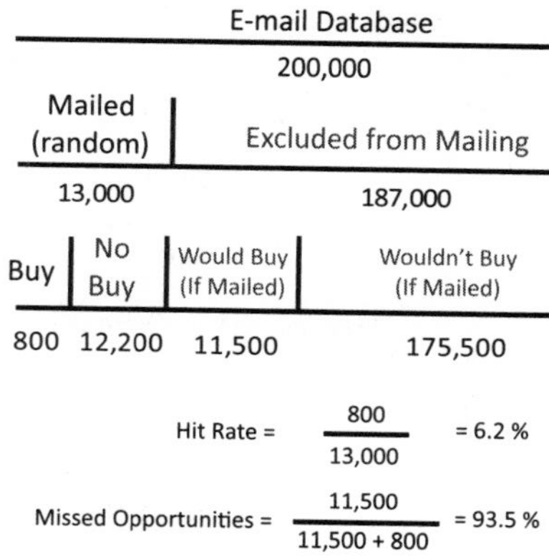

		E-mail Database	
		200,000	
Mailed (random)		Excluded from Mailing	
13,000		187,000	
Buy / No Buy	Would Buy (If Mailed)	Wouldn't Buy (If Mailed)	
800 12,200	11,500	175,500	

$$\text{Hit Rate} = \frac{800}{13,000} = 6.2\,\%$$

$$\text{Missed Opportunities} = \frac{11,500}{11,500 + 800} = 93.5\,\%$$

FIGURE 4-4 Method Two of Targeting: Groupon markets to 13,000 names selected by a targeting model. The expected hit rate of those targeted (2,400/13,000 = 18.5%) is three times that of the average hit rate (6.2% from Figure 4-3).

TARGETED SELECTION OF 13,000

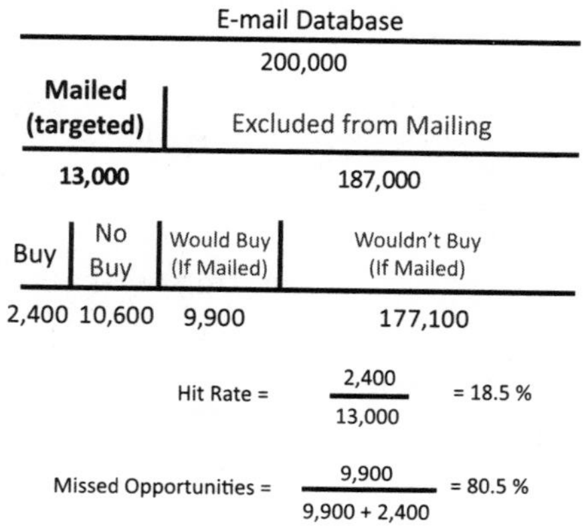

		E-mail Database	
		200,000	
Mailed (targeted)		Excluded from Mailing	
13,000		187,000	
Buy / No Buy	Would Buy (If Mailed)	Wouldn't Buy (If Mailed)	
2,400 10,600	9,900	177,100	

$$\text{Hit Rate} = \frac{2,400}{13,000} = 18.5\,\%$$

$$\text{Missed Opportunities} = \frac{9,900}{9,900 + 2,400} = 80.5\,\%$$

FIGURE 4-5 Method Three of Targeting: Groupon markets to 26,000 names selected by a targeting model. Expanding the list of targets increases the number of coupons sold (from 2,400 to 2,830) but reduces the hit rate (2,830/26,000 = 10.9%) compared to Figure 4-4 due to the law of diminishing returns.

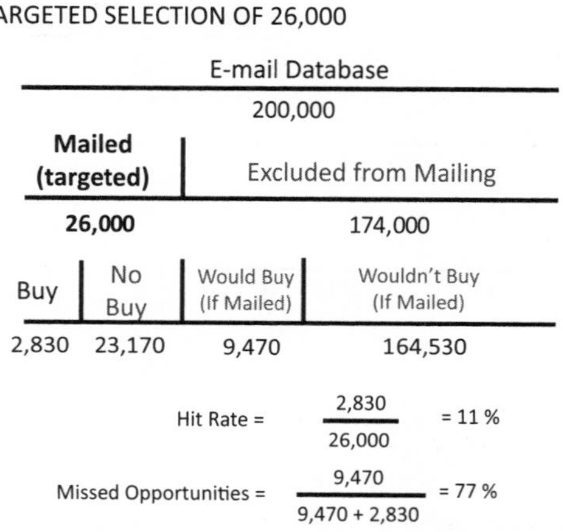

TARGETED SELECTION OF 26,000

	E-mail Database		
	200,000		
Mailed (targeted)		Excluded from Mailing	
26,000		174,000	
Buy / No Buy	Would Buy (If Mailed)	Wouldn't Buy (If Mailed)	
2,830 23,170	9,470	164,530	

$$\text{Hit Rate} = \frac{2,830}{26,000} = 11\,\%$$

$$\text{Missed Opportunities} = \frac{9,470}{9,470 + 2,830} = 77\,\%$$

Using the framework from Chapter 4 of *Numbers Rule Your World*, we can infer Groupon's strategy:

> The coupon vendor has an incentive to minimize false negatives while tolerating false positives. A missed sale drops the bottom line directly whereas false alarms may upset a few subscribers.

Bear in mind that Groupon subscribers knowingly ask to receive daily e-mails—people like the *New York Times* reviewer David Pogue who enjoyed "giddiness," "thrill," "serendipity," and "exclusivity." Thus, putting self-interest above all, Groupon should aggressively expand its reach. Ironically, that means the managers should order less targeting, not more!

5. Wanted: Newbies

If Groupon takes care of itself first, it would moderate the usage of targeting. *Targeting* is an act of restricting the scope, of cropping the roster of subscribers eligible for a particular deal. By shunting people with a lower chance of purchase, the statistical models elevate the rate of response in the short list. But a better rate does not always equal more units sold. Say the uber-algorithm lifts the effectiveness from 10,000 sales off 100,000 e-mails to 8,000 sales off 50,000 e-mails. As the hit rate expands from 10 percent to 16 percent, Groupon sells 2,000 fewer coupons. A win for the modelers is a loss for the sales force. Since missed opportunities are expensive to Groupon, which collects roughly half of the coupon value from merchants, using targeting is shooting itself in the foot. Something else has to constrain Groupon's profit motive, and justify its enthusiasm for targeting technologies.

That impetus comes from the merchants who comprehend the "Grouponomics" of Chapter 3. They realize they must strike a balance between two types of coupon buyers, the free rider, who effects a loss, and the newbie, who generates an incremental profit.

Merchants value targeting, but not the kind favorable to Groupon. Pretend you are the owner of a neighborhood pizza parlor. Peter is a regular customer who drops by every Thursday after picking up his son from basketball practice. David works as an instructor at the gym down the block but for some reason, he has never eaten at your shop in the three years he's lived in the neighborhood. Is Peter or David more likely to prepay for a Groupon? A targeting model that maximizes Groupon's revenues would direct coupons to Peter rather than to David. This result displeases you, the shop owner, because you want to entice David to try your food, and you expect Peter and his son to show up each Thursday even if Groupon didn't exist.

Merchants view targeting models in a different light. Instead of ranking Groupon subscribers by the likelihood to purchase coupons, retailers want the algorithms to cherry-pick the newbies while fencing off the free riders. These models should assign a rating that measures the probability of being a newbie. Compare the merchant's perspective with Groupon's perspective in Figure 4-6. The outcomes of the targeting algorithms differ because free riders are, in general, more likely to purchase coupons.

To a Groupon merchant, a false-positive is a coupon delivered to a regular customer, and a false negative is a potential new customer not targeted by Groupon. The former inflicts a direct loss of revenues while the latter represents a missed opportunity. The merchant worries about both types of mistakes, and in the case we studied in Chapter 3, the sweet spot is where 70 percent of redemptions come from first-timers.

Earlier, we wondered why so many coupons in Augustine Fou's inbox miss their mark. Here is another reason: merchants prefer sending deals to those unaware of their

---

FIGURE 4-6 Conflicting Objectives of Targeting: (a) Groupon maximizes its revenues by directing deals to the most likely purchasers but those segments are also likely to include a higher proportion of free riders. (b) Merchants, by contrast, optimize the profitability of Groupon promotions by targeting segments of newbies.

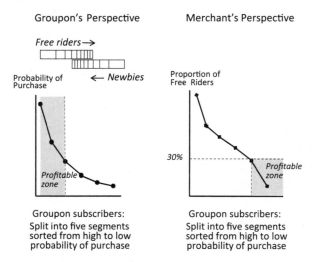

Groupon's Perspective

Merchant's Perspective

Free riders →

Probability of Purchase ← Newbies

Proportion of Free Riders

Profitable zone

30%

Profitable zone

Groupon subscribers:
Split into five segments
sorted from high to low
probability of purchase

Groupon subscribers:
Split into five segments
sorted from high to low
probability of purchase

goods or services, but customers prefer hearing from merchants with whom they are already familiar. As Pogue noted, we deem most relevant the coupons for "all the stuff we'd buy anyway." The harder Groupon tries to move us out of our comfort zone, the more we think its targeting machine is misfiring. A targeting algorithm cannot satisfy two conflicting objectives at once.

6. The Targeting of Groupons

On November 4, 2011, Groupon proved that its IPO baby was not ugly, as Andrew Mason, its irreverent founder, now a thirty-something billionaire, had predicted in a staff memo of August 2011. Evidently, the win-win-win story won over investors, just as it had impressed Felix Salmon, David Pogue, and other seasoned analysts.

The beauty of Groupon's story is its simplicity: Who doesn't like a deal? But figuring out its business model is not trivial as the daily-deals service has never made a profit in 36 months from inception to IPO. High-tech IPOs present an ideal setting to practice NUMBERSENSE. Founders and funders are selling a vision built on scientific foundations. Google showed off its trophy PageRank algorithm for Web search, and wanted to share its dream of capturing and organizing all information. Amazon started its life as a fledgling online bookseller, and promised to become the world's largest retailer. Common to these business propositions is the element of science fiction—a short commercial history paired with an elaborate fantasy of domination. How do we tell if Groupon is the next Google, or the next Webvan?

NUMBERSENSE is not taking numbers at face value. NUMBERSENSE is the ability to relate numbers here to numbers there, to separate the credible from the chimerical. It means drawing the dividing line between science hour and story time. With a bit of quantitative thinking, we turned up a host of surprising insights about Groupon's business.

We diagnose a short leg in the win-win-win story: Consumers clearly win, Groupon also wins, but not all merchants win. Unexpectedly, retailers may lose business even as their stores swell with customers. We're talking about bad news on the top line, not just the bottom line: more customers producing lower total revenues than would have been earned without Groupon. If daily deals are a form of free advertising for retailers, then "free" arrives with an asterisk. In every case, the merchant funds the consumer's saving and pays the vendor a share of revenues.

Targeting technology is one tool that can strengthen the economics of a Groupon merchant. But the punditry fails to comprehend how. Targeting as described is not so much concerned with sending more relevant deals to subscribers; it works by directing coupons to profitable segments of customers, away from the free riders and toward the first-timers. Groupon's two-sided marketplace behaves differently from typical businesses—the more attractive for the consumer, the more draining for the merchant. If the commenters had explored the mathematics of targeting, they would have learned that targeting is error-prone; and models that ace statistical standards of accuracy still make plenty of incorrect predictions. It is, therefore, no accident to find our inboxes flush with deranged deals. If Groupon tunes its targeting machinery wisely, its merchants will run more cost-effective promotions. A side effect of such success is subscribers purchasing fewer coupons and Groupon collecting less revenues. Nonetheless, Groupon investors who realize the complexity of its business model should embrace this seemingly self-defeating strategy of deal targeting.

7. Growing Pains

It was May 2011, when I read Felix Salmon's "Grouponomics" post, and responded with a blog titled "Grouponomics and the Power of Counterfactual Thinking." That essay sowed the seeds of Chapters 3 and 4 of this book.

Six months later, the daily-deals company impressed investors and confounded critics, ending its first trading day at a price of $26, about 30 percent above the IPO price. The floatation was one of the biggest in history, second only to Google's in 2004.

The comparison with Google, however, ends there. While Google's advertising business, based on its ground-breaking algorithm for Web search, produced over $40 billion of revenues in 2012, Groupon stuttered early and often. Hardly one week past its first year as a public company, GRPN stock had sunk by 90 percent to $2.60.

By March 2013, Andrew Mason's tenure as CEO ended abruptly. In his final earnings call with investors, Mason reported that fourth-quarter revenues in 2012 grew 30 percent relative to 2011. Nonetheless, the business of selling daily deals slowed from the third to the fourth quarter, domestically as well as internationally. The fourth quarter is make-and-break for the full year for any retail company.

Groupon managers touted an initiative to sell goods directly to consumers. This Groupon Goods division contributed $225 million in sales during the fourth quarter of 2012, but with a ghastly gross margin of 3 percent. The struggling retailer has invaded a field dominated by Amazon. (Amazon's gross margin has exceeded 20 percent for the past five years.)

Meanwhile, Amazon's foray into daily deals through a $175-million stake in LivingSocial had grounded. The leader of online retailing wrote off substantively the entire investment in October 2012, when it reported a first net loss in four years. In February 2013, LivingSocial raised $110 million from existing investors to stay alive. Its CEO admitted that "this was a down round": The second largest daily-deals company was valued at $1.5 billion.

Nearly two years before, the premier technology blog, *TechCrunch* had the scoop on LivingSocial's valuation: $2.9 billion.

Why Do Marketers Send You Mixed Messages?

The mass retailer Target, with the famed red bull's-eye logo, made the front page of *New York Times Magazine* in 2012 with an eye-catching—to others, appalling—application of customer targeting. The journalist Charles Duhigg describes the working of a statistical model that predicts if a female customer is in the second trimester of pregnancy:

> Take a fictional Target shopper named Jenny Ward, who is 23, lives in Atlanta and in March bought cocoa butter lotion, a purse large enough to double as a diaper bag, zinc and magnesium supplements, and a bright blue rug. There's, say, an 87 percent chance that she's pregnant and that her delivery date is sometime in late August . . . [The marketers at Target] know that if she receives a coupon via e-mail, it will most likely cue her to buy online. They know that if she receives an ad in the mail on Friday, she frequently uses it on a weekend trip to the store. And they know that if they reward her with a printed receipt that entitles her to a free cup of Starbucks coffee, she'll use it when she comes back again.

Marketers have marked pregnancy as one of the few life events that cause women to alter their shopping habits, and if Target beats other retailers to the chase—by jumping in front of the barrage of deals that bombard these women once the birth records become public—Target could steal market share from its competitors. According to Duhigg's informant, Target achieved a remarkable spurt in the sales of baby products after deploying a marketing program to reach women predicted to be pregnant.

This type of predictive technology benefits from Big Data—the construction of gargantuan databases that record every trivial interaction between a customer and a business. Chris Anderson, the former editor-in-chief of *Wired* magazine, once argued that when data become plentiful, every detail is exposed, nothing needs explaining, and theory is passé. Are we entering such a world? Is this world as scary as it sounds? As more and more companies invest in targeting machinery, it is imperative that we learn what they are doing, and how they are doing it. How accurate are Target's predictions? Something about information technologies causes reporters to lose their bearing, and so it is with the practice of customer targeting: Does the substance justify the hype?

1. How the XXL Purse Gave You Away

Direct marketers have been using targeting models for many decades, before the arrival of the Internet. Banks such as Citibank, Capital One, and American Express send preapproved applications for credit cards based on targeting models—not everyone's mailbox is stuffed with their unsolicited mail, in case you're wondering. Mail-order businesses vary the size and contents of their numerous catalogs according to predicted customer types; Williams-Sonoma, for instance, saves 20 percent off the cost of postage by mailing thinner catalogs to selective segments of customers. More recently, online retailers such as Netflix and Amazon make "personalized" recommendations of movies and books. Google analyzes the contents of e-mails in order to place more relevant advertisements next to them. Casino operators such as Harrah's offer special rewards to customers chosen based on their spending habits, which are tracked by loyalty cards.

Target wanted to boost sales to the key demographic group of new mothers. A team of data scientists went to work. They created a predictive model which assigns every shopper a "pregnancy rating," interpreted as the chance that she is carrying a baby. The rating formula considers how much she's recently spent on 25 carefully curated products. (See Figure 5-1 depicting our example mom-to-be Jenny Ward.) The modelers discovered that the purchase of these items precedes childbirth.

One popular method of targeting is *market-basket analysis*. Imagine Target taking a snapshot of your shopping basket each time you walk out of a store; now stash the pictures

---

FIGURE 5-1 The Mass Retailer Target Uses Prior Purchases to Predict Future Purchases: The predictive model produces a "score" which rates the chance that the shopper would be pregnant.

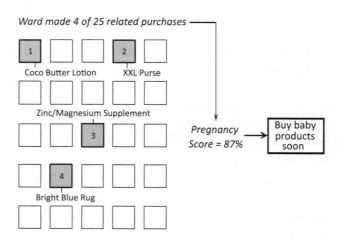

to make a flip book. Target can play back the sequence of everything you've ever purchased from them. By reviewing the moving images of thousands of customers, the modelers discover recurring patterns: For example, many shoppers who purchased an extra-large purse will eventually buy a baby crib.

With a modest investment in computing infrastructure, any retailer can build profiles of its customers. The analysts start by tracing your past purchases, but they know a whole lot more about you:

- How long have you been a customer?
- How much have you spent in total?
- How much have you spent recently?
- What's the average bill?
- Is your expenditure trending up or down?
- How long has it been since your last purchase?
- How wide a range of products have you purchased?
- Do you buy off-the-shelf or customized products?
- Are you an early adopter of new releases?
- How many calls for service have you made?
- Do you read marketing e-mails?
- Do you make use of coupons?
- Are you price-sensitive?
- How satisfied are you?

This list goes on and on.

Absorbed by Duhigg's engaging narrative, you may have over-looked the detail that the market-basket analysis described above cannot make predictions! To certify the link between the 25 products and baby goods, the analysts produce a stack of flip books; but for any woman who fits the shopping pattern, it's too late to win her business. The real goal of the analysis is to define interesting groups of customers: Jenny Ward "looks like" someone from these groups—except baby goods haven't appeared on her shopping list yet. Thus a prediction is borne. You will be reminded of how Miranda Priestly comes up with fashion tips for Taylor Nitiolex in Chapter 4. The "look-alike" principle is fundamental to all predictive models.

Look-alike modeling based on past customer transactions is very powerful but there's a catch. Unless you shop regularly at Target, they have limited direct data about your purchasing habits. If you make only a few purchases a year, it's unlikely you've bought enough of those 25 products to matter, and then it's virtually impossible to guess whether you're pregnant. How many Jenny Wards are out there? We're talking about a customer who shops at Target on many weekends, spends money across a variety of product categories, redeems coupons at Target's online store, reads marketing e-mails, and loves the Starbucks offer. Ward is probably one of Target's best customers, even before she got pregnant. But consider this: A targeting model really shouldn't focus on the Jenny Wards since they'll drop by a Target soon anyway. The infrequent or first-time customers are the ones who pose the staunchest marketing challenges, and yet for whom the store knows little of relevance.

Say Amazon wants you to purchase *Numbersense* from them but you like to stock your bookshelves from your favorite independent bookstore. Having mined the customer data, Amazon learns that people who have read *Fooled by Randomness* and *Freakonomics* in the prior six months are more likely to buy *Numbersense* now. You belong to this segment of people, but Amazon wouldn't know it since you picked up the related books from the indie bookshop. How does the online retail giant locate you?

The targeting machine takes another pass through the purchasing habits of probable *Numbersense* readers, hoping to uncover common traits among them:

- Do they come from a particular age group?
- Do they live in certain parts of the country?
- Are they male or female?
- What magazines do they subscribe to?

- Are they heavy users of the Web?
- How frequently do they order things using their cell-phones?

Eventually, profiles of several types of customers emerge as targets. One possible set of buyers may be college-educated people over 40 who hold managerial jobs and live in one of the top 25 metropolitan areas. If Amazon places you in this group, the next time you visit their store, they will recommend my book to you. The notion of "one-to-one" marketing is horribly overhyped if you realize that most merchants don't know enough about you to make truly personal offers.

2. What Companies Know about You

Retailers like Amazon have two ways to guess what you'd buy next. If you are a loyal customer, they flip through your old shopping receipts looking for clues. Otherwise, they bind you to some regular customers who "look like" you. This linkage is enabled by *proxy data*, such as age, income, magazine subscriptions, and ownership of pets.

The most direct way to learn your shopping habits is through *loyalty cards*. Retailers, in effect, pay for your personal data with rebates, gifts, and other goodies. When Amazon issues a credit card through Chase Bank, customers earn three points per dollar of purchases from Amazon and one point for most other charges. The triple-point incentive encourages you to direct the bulk of your expenditures to the Amazon card, which ensures that Amazon has a more-or-less unimpeded view of your spending patterns.

For the infrequent customer, retailers must rely on proxy data. The amount of data out there is staggering. Companies, such as InfoUSA, Experian, and Epsilon, own galactic databases that cover more than 75 percent of all U.S. households. Collecting and selling the data is their business. They sell:

- Demographic data, like gender, age, ethnicity, education, and income
- Neighborhood data, including the proportion of people around you who own homes, or the proportion of your neighbors whose daily commute exceeds 60 minutes
- Consumption data, such as how much you spend on ice cream or from home shopping channels
- Lifestyle data, including when you moved residences and when you got married

In recent years, data about online or mobile usage are being compiled and marketed by start-ups like BlueKai and eXelate. The scale of these operations is immense: They keep tags on over 100 million Web surfers every month, just in the United States alone. These data exchanges occupy a corner of the Big Data ecosystem. Big Data became capitalized in the 2010s, heralded as the "Next Big Thing" in high tech, succeeding other waves such as social media, broadband, and Web search. Accel Partners, one of the legendary names in Silicon Valley, launched a $100-million venture-capital fund dedicated to supporting Big Data start-ups. When Facebook finally unveiled its IPO plan in early 2012, analysts regarded its lofty valuation—up to $100 billion—as "cashing in on personal data." The premier social networking service is presumably the largest repository of proxy data that can be harnessed by targeting models.

While users of Facebook, LinkedIn, Twitter, and similar services voluntarily make their information public, some Big Data companies deploy clandestine methods to harvest personal data. A series of controversies has shone a light on these practices. In December 2011, a software programmer named Trevor Eckhart demonstrated how an app by Carrier IQ, deeply embedded in most smartphones, was beaming data back to its servers, including the contents of personal text messages, without permission from users. Not a few months passed when another developer, Arun Thampi, discovered that Path, an upstart alternative to Facebook, was secretly uploading the address books of iPhone users. On further examination, a host of other app developers were committing the same offense, an explicit violation of the guidelines set down by Apple, the gatekeeper of iPhone apps, in which developers agree to obtain prior consent from users.

Whether we like it or not, these technology companies are turning their figurative webcams on all of us. And we have been warned; in 2009, Eric Schmidt, then CEO of Google, quipped: "If you have something that you don't want anyone to know, maybe you shouldn't be doing it in the first place." This comment came from the captain of a fleet of cars that circulate in the country snapping images for the Google Maps with Street View service. When authorities investigated those roving vehicles, they discovered much more than photos; e-mails, passwords, search histories, and more were being plucked out of the ether.

In 2008, Chris Anderson, former editor-in-chief of *Wired* magazine and author of *The Long Tail*, made a bold prediction about Big Data years before the irrepressible hype arose. In an article titled "The End of Theory," Anderson asserts that data would become so plentiful that everyone and everything would be fully detailed and reality revealed in all its splendor at any level of precision and there would no longer be a need to create models that bastardize reality. In Anderson's own words:

> This is a world where massive amounts of data and applied mathematics replace every other tool that might be brought to bear. Out with every theory of human behavior, from linguistics to sociology. Forget taxonomy, ontology, and psychology. Who knows why people do what they do? The point is they do it, and we can track and measure it with unprecedented fidelity. With enough data, the numbers speak for themselves.

Such a provocative vision deserves unpacking—if only because our popular press is capable of cult worship when it gazes at information technologies. How accurate are statistical models based on correlations? To what extent does the volume of data affect "fidelity"?

3. The Science of Sending Mixed Messages

Charles Duhigg's ode to targeting models reached a signal moment when Target marketers told an unsuspecting Dad that his daughter was pregnant even before she informed her parents! What originally appeared to be a case of irresponsible targeting of innocent, young girls turns into a triumphant tale of applied statistics. But how much can we trust models based on correlations?

Let's say that at any time 10 percent of the women on Target's roster of customers are pregnant. A targeting algorithm will label 20 percent of the female shoppers as (probably) pregnant, and if the model is Miranda Priestly–caliber, about 6 of the 20 percent would be accurately predicted. Thirty percent of those who are predicted pregnant would turn out to be pregnant, and by this measure of accuracy (technically called the *positive predictive value*), the model disappoints. Nonetheless, statisticians rate this model top drawer, because 30 percent is three times the incidence of pregnancy at large: Customers picked by the algorithm are three times more likely to be pregnant than the average female shopper. This is the "lift" that experts use to quantify the value of the modeling. Still, the impressive model fails to identify 40 percent (4 out of 10) of pregnant women, while 14 percent of female shoppers (20 minus 6) will, if they pay attention to Target's marketing materials, wonder when the hip mass retailer turned into Babies "R" Us. Figure 5-2 explains how to compute the various rates mentioned above.

Duhigg presented a puzzle: If the targeting technology reliably identifies a woman's pregnancy, in some cases, even

FIGURE 5-2 Evaluating a Predictive Model: For every 100 female shoppers, 10 are actually pregnant. In order to find those 10, the model selects 20 as most likely to be pregnant. The name of the game is to find the 10 targets with as short a list of targets as possible.

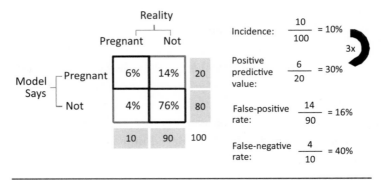

before her family noticing, why do Target's marketers dilute their message by mixing in randomly selected, unrelated products? Why the coyness in deploying such a potent technology? Duhigg suggested that this evasive action counteracts the creepy precision of Target's predictive model. An unnamed Target executive elaborated:

> We'd put an ad for a lawn mower next to diapers. We'd put a coupon for wineglasses next to infant clothes . . . as long as a pregnant woman thinks she hasn't been spied on, she'll use the coupons.

But now, ask yourself which scenario is the more embarrassing? To send a brochure filled with baby products to a pregnant woman who does not expect Target to know about the impending happy occasion or to send the same brochure to a customer who isn't pregnant at all? The second scenario stinks worse. And we know from the analysis above that 70 percent of those receiving brochures would not actually be pregnant! Could it be that mixing in random products serves to cover up the many misfires of targeting models?

4. Is Big Data a Savior?

Statistically speaking, the best predictive models are gems. Even so, most targeted customers are false positives. This unfortunate outcome is itself predictable because business executives hate missing sales more than they fear the tongue-lashing from customers offended by incoherent marketing. Would the advent of Big Data save the day?

Let's explore how you decided to pick up *Numbersense*. Perhaps you saw the book displayed in a shop window, and the cover design caught your fancy. Perhaps you enjoyed my previous book on statistical thinking. Perhaps it was your birthday, and you got yourself a little treat. Perhaps on the first day of every month, you had the habit of purchasing a new book from your local bookstore. Perhaps a coworker gave a glowing review of the book that morning. Perhaps you rarely read business books, but you picked this up on a whim. Perhaps you are a loyal reader of my blog. Perhaps your spouse teaches math. Curiosity, joy, friendship, peer pressure, habits, gullibility, impulsiveness, fads . . . these are all plausible reasons for buying *Numbersense*.

Now, ask yourself if any of the following caused you to buy:

- You are middle-aged.
- You have a college degree.
- You manage people at work.
- You live in a city.

Do you see that these cannot be the true reasons for buying *Numbersense*? The statistics may show that most buyers are city dwellers, but no one paid for the book *because* he or she prefers urban living. A touch of counterfactual thinking clears the fog: If someone had raised a family in the suburbs, in all likelihood, he or she would still have purchased the book. And yet, the standard targeting models devour such data as age, education, occupation, and geography. Target's algorithm uses the pattern of past purchases, which is also an indicator, not a cause, of future purchases. It is oblivious to the intangible quantities that more directly affect one's behavior: trust, peer influence, habitualness, and so on.

Unfortunately, the true causes for buying defy simple measurement, if they can be measured at all. Statistical models in the social sciences rely on correlations, generally not causes, of our behavior. It is inevitable that such models of reality do not capture reality well. This explains the excess of false positives and false negatives.

A statistical model is nothing like Newton's model of gravity, in which the downward force causes the apple to fall from the tree, be it yesterday, today, or tomorrow. Real-life correlations, however, are far from consistent. If you are carrying a green umbrella today, one can't be sure the next umbrella you purchase would also be green. A model that ignores cause–effect relationships cannot attain the status of a model in the physical sciences. This is a structural limitation that no amount of data—not even Big Data—can surmount.

On the contrary, an abundance of data tends to invoke a trust in correlations that is undeserved and errant. In his best seller *The Black Swan,* the economist Nassim Taleb warns readers not to dispel the possibility of a black swan, regardless of how many white swans they see. Big Data fights Black Swan; Black Swan wins.

Statisticians devote much labor to building more realistic cause–effect structures into social-science models. These more advanced structures tend to mimic Figure 5-3b. The conceit is to ask the algorithms to do what humans can't—to tease out the real causes like fads and impulsiveness. These elements are called *latent factors* because they cannot be observed directly. The modeler honestly can have no idea what the hidden factors are measuring, so he makes assumptions or interpretations, neither of which can be verified. He may even decide to leave the latent factors unexplained. This trick evidently does not resolve the structural problem but as I discussed in *Numbers Rule Your World*, imperfection in a statistical model is permitted so long as the model is able to produce additional insights about the mysterious world.

We have good reason to believe that such causal structures are unstable anyway. In the last few decades, behavioral psychologists, using ingenious experimentation, have discovered that our judgment is easily swayed by *priming effects*. In one such setup, designed by business professors Chen-Bo Zhong and Katie Liljenquist, test subjects were asked to copy a story. One group copied a story about *sabotaging* a coworker while the other group wrote out a story about *helping* a coworker. After finishing, all participants filled out a survey in which they rated the desirability of a range of household

---

FIGURE 5-3 Latent Factors in Modeling Consumer Behavior: (a) A simple model that analyzes past behavior is limited by not capturing the true causes of the behavior. (b) In a more ambitious structural model, observed indicators—such as past purchases—are used to estimate latent factors—such as desire for knowledge, peer influence, impulsiveness, and gullibility toward marketing—which are posited to affect the current purchasing decision directly. The latent factors are not measurable, even as they reflect the causal beliefs of the modeler.

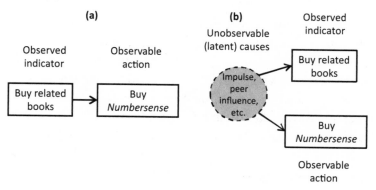

products. Since tedious copying is unrelated to shopping, we should expect both groups to express similar views about various products. Surprise! Both groups did give similar ratings to a subset of products, such as Post-it notes and Energizer batteries. For cleansing products specifically, such as Crest toothpaste and Tide detergent, those who copied a story about sabotaging a coworker found them significantly more desirable than those who copied a story about helping a coworker. In such experiments, almost all test subjects, when questioned afterward, rejected the idea that they could have been affected by the priming activity. So it appears that the researchers manipulated (caused) people to want cleansing products by priming them with an irrelevant activity.

Daniel Kahneman, a professor at Princeton University, and one of the leading thinkers in behavioral psychology, documents the ground-breaking research on priming effects and other unexpected biases in decision-making in his magnum opus, *Thinking, Fast and Slow*. (Kahneman's most productive collaborator was the late Amos Tversky.) Consider the implication of priming. So many things could predispose one's behavior. Multiple priming factors may be in effect simultaneously. The effect may only last for some unknown amount of time. Even after the effect has been demonstrated, people would not believe they have been affected. The results from various experiments threaten the search for stable, logical, causal structures that explain our decisions. The absence of explanations consigns statisticians to modeling correlations, an activity that is inherently prone to errors not curable by data infusion.

Perhaps the views of Chris Anderson, who lives in California, have been shaped by conversations with people in the high-tech industry, an arena in which model errors have light consequences. If Google's PageRank algorithm fails to find the truly most relevant web pages for your search, the company suffers no real harm—would you even notice? If your Netflix recommendations are crappy, you'd just ignore them.

Groupon bombards digital marketer Augustine Fou with shots of irrelevant deals, but he's not about to grumble about something he gets for free. "With enough data, the numbers speak for themselves," Anderson asserted in 2008. The unspoken truth is that most predictions made by these correlational models are wrong. It's not a matter of smarts or skills. It is just hopeless to distill the kaleidoscope of human behavior into a set of equations. That's why Big Data will not spell the end of theory. Any statistical model includes some assumed theory, a topic we'll delve into in the next two chapters.

PART 3

ECONOMIC DATA

Are They New Jobs
If No One Can Apply?

February 2, 2010. Groundhog Day. People assembled at a morning festival to greet the furry squirrel and if it came out of hiding, spring would arrive early that year. Little did we know there were at least 20 prognosticators; 13 of them predicted a shortened winter. The famous one, Punxsutawney Phil, which is featured in the Bill Murray movie, was one of the seven dissenters.

Perhaps Punxsutawney merited the limelight, for 10 days later, on February 13th, the extraordinary happened: There was snow cover in every state of the union, except Hawaii. Winter was clearly not taking leave. The Northeast corridor endured two successive weekends of terrible blizzards, the first one on February 5 and 6, and the second from February 10 to 13. Hundreds of thousands of people in Washington, D.C., lost power. Museums, monuments, and the White House were closed to visitors. The U.S. Postal Service did not deliver mail for the first time in three decades. Thousands of flights were cancelled. Many locales received two to three feet of snow during the first storm, and an additional one to two feet during the second, dubbed "Snowmageddon." Concurrent with "Snowmageddon," the Deep South encountered a rare storm, delivering six inches of snow to Louisiana, while also bringing the white powder to Florida. Dallas, Texas, recorded 11.2 inches of snow in one day, an all-time record.

Those were only two of three gigantic storms of February 2010. The third, called "Snowicane," appeared on February 25th. By the end of the month, total snowfall had broken all-time records in: Baltimore, Maryland (49.1 inches); Washington, D.C. (46.1 inches); New York City-Central Park (36.9 inches); New York City-LaGuardia (29.1 inches); Pittsburgh, Pennsylvania (48.7 inches); and other places. Eventually, new season records were written in Washington, D.C.; Baltimore, Maryland; Philadelphia, Pennsylvania; Wilmington, Delaware; and Atlantic City, New Jersey.

The Friday after "Snowicane," the Labor Department was scheduled to release its monthly *Employment Situation Report*. Economist Mark Rogers, who spent 19 years at the Federal Reserve Bank of Atlanta, calls it "the most closely followed economic report on earth." This makes the gain or loss of jobs and the unemployment rate two of the most essential economic indicators in the world. When the Great Recession struck in December 2007, they also became the scariest statistics ever. More than 8 million jobs have evaporated in the United States since. The depth of this human tragedy is poignantly displayed in Figure 6-1. The popular blog, *Business Insider*, dubbed this the "scariest job chart ever." If you were a policymaker, this chart should be making you sweat in your bed. During the last economic downturn of 2001, it took nearly four years for the jobs market to trudge its way back to the pre-recession state. Given the anemic recovery path as of 2012, economist Dean Baker estimated the nation's employment would not return to full health until 2028! Full health requires more than offsetting the job losses: Because the U.S. population has grown in the meantime, we need even more jobs than before just to keep the same proportion of people employed. It's a case of trying to run up a down escalator.

FIGURE 6-1 The Scariest Jobs Chart: Every post–World War II recession led to great losses in jobs, followed by eventual full recovery. For simplicity, I present the fall and rise as two straight lines. [*Source*: Adapted from the *Calculated Risk* blog.]

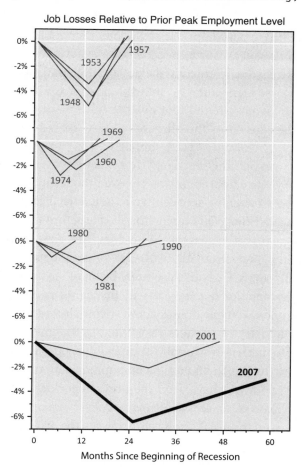

1. A Snow Job

By early 2010, after 24 straight months of job losses, everyone had grown tired of calling the turning point. Forecasters had been embarrassed by the unpredictable economic currents. It had been more than ten months since the U.S. Federal Reserve Chairman, Ben Bernanke, grabbed headlines in a *60 Minutes* broadcast, invoking the words "green shoots." For many American workers, spring is the friend who's always planning to visit next year. Worse, the February snowstorms looked to push hope back by yet another month. Larry Summers, President Obama's top economic advisor with his impeccable academic credentials, apparently confirmed everyone's worst fears when he told the host of CNBC's *Fast Money* that people should blot out the soon-to-be-updated jobs number, because "the blizzards . . . are likely to distort the statistics." Summers even supplied a tip: "In past blizzards," he advised the viewers, "those statistics have been distorted by one-hundred to two-hundred thousand jobs."

Summers's warning shot sent veteran financial columnist John Crudele reeling. On March 4th, the day before the Labor Department report, he cautioned his *New York Post* readers: "Expect Snow Job from White House on Jobs Report." He sarcastically applauded an "absolutely brilliant" move by the executive branch, a master class in managing perception. The pre-announcement non-announcement shifted the goal posts, as industry forecasters responded to Summers's cue, like metal to magnet, by sharply lifting their expectation of job losses from 20,000 to 68,000. As an administration insider, Summers just might have seen the preliminary data. When the number turned up as 36,000, market observers responded positively, even though it significantly missed the original estimate, the one without the Summers-induced correction.

Crudele was sure that Summers was wrong. The February snowstorms could not affect the jobs report in any meaningful way. His argument wasn't solely based on speculating about other people's motives. Having written dozens of columns on employment figures, he knew—in impressive detail for a *New York Post* writer—how the Bureau of Labor Statistics (BLS) counts jobs. One element of NUMBERSENSE is learning the origin of data, and here we have a nice illustration of it.

Each month's employment situation report comes with a section called "Technical Note." In it, the Labor Department describes the two surveys used to measure the health of the nation's work force:

- The payroll survey, properly known as *Current Employment Statistics* (*CES*), gathers data from 150,000 businesses and government agencies.
- The household survey, or *Current Population Survey* (*CPS*), consists of interviews with 60,000 households selected to represent the entire nation.

Hypothetically, the February blizzards could foul up survey results in two ways: Some people could not work due to the weather, or some employers could not return the survey on time.

There is no doubt that the snow caused absences from work. You'd think the jobs number should have been affected, but you'd be wrong. As Crudele explained, it depends on the rules of counting. The payroll survey tallies every job for which someone receives pay during the pay period that includes the twelfth day of the month. February 12 was a Friday in 2010. (See Figure 6-2.) As most pay checks are issued either bimonthly, weekly, or monthly, the reference week for the payroll survey was either February 1 to 12, February 8 to 12, or February 1 to 26. Employers are requested to count any employee who gets paid for one or more hours of work during the reference week. Since few people were kept out of work for more than a few days, those brief absences could hardly have skewed the statistics.

Meanwhile, the household survey counts every person who works for at least one hour during the calendar week containing the twelfth day of the month, which meant February 8 to 12. BLS, however, does not require workers to be present at work in order to count them as "employed." They maintain a category called "with a job, not at work," which includes people who cannot work due to bad weather.

FIGURE 6-2 Snow Days of February 2010: February 12 was a Friday in 2010. If you are paid twice monthly, then the employer would report your status for the first two weeks of the month to the CES survey. The calendar shows the snow days relative to the reference week.

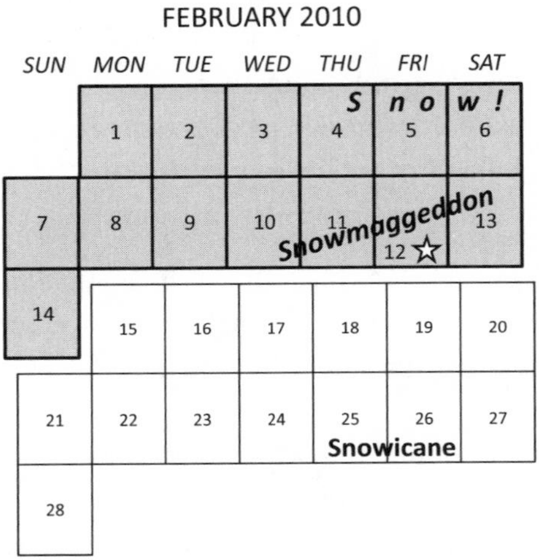

FEBRUARY 2010

SUN	MON	TUE	WED	THU	FRI	SAT
			S	**n o**	**w !**	
	1	2	3	4	5	6
7	8	9	10	11 Snowmaggeddon	12 ☆	13
14	15	16	17	18	19	20
21	22	23	24	25	26 Snowicane	27
28						

What about the theory that the blizzards prevented some businesses from returning the survey? Imagine a questionnaire addressed to Ann's Scones and Jams Co. In January, the corner bakery employed a staff of 10 people. On February 9th, Ann slid on black ice and fell heavily, breaking both legs. Consulting her almanac while bedridden, she guessed that the remainder of February would be lousy. She decided to take the month off, and didn't bother to fill out the payroll survey.

The survey analyst faces a missing-data problem. One common solution is *zero imputation*, where the analyst substitutes every blank with a zero. Doing so effectively treats all businesses that did not return the survey as having ceased operations. This assumption is clearly flawed, as it takes *too many* real jobs out of the count. Statisticians have a cautionary saying: Absence of evidence is not evidence of absence. Type "zero imputation" into a search engine, though, and you'd be surprised how frequently it is implemented in a variety of settings.

A different remedy is *mean imputation*. Here, the analyst assumes that the non-responders would have given the same answers as the responders. This is another brave assumption, perhaps a tad less wanton. It removes *too few* real jobs out of the count.

BLS statisticians do not jump to the conclusion that Ann's store has forever shuttered, taking 10 jobs out of the economy. Instead, they employ a form of mean imputation to deal with business deaths. (I return to this arcane issue at the end of the chapter.)

If we know about these generally lenient rules for counting jobs and employed people, we can tell that a few days of inclement weather could not have eliminated 100,000 to 200,000 jobs. If we forget, there is Crudele to remind us once every few weeks in the *New York Post*.

2. To Season or Not to Season

On the first Friday of each month, the Department of Labor releases the nation's jobs report. The next day, *New York Post* financial columnist John Crudele guides his readers to "the truth." For example, on February 3, 2012, the media welcomed an announcement of 243,000 new jobs, much above the consensus forecast of economists. The following morning, Crudele branded the report "a ruse." He urged his readers to look up the raw data. "In truth," he explained, "Labor's survey of companies found that 2,689,000 jobs had disappeared in January . . . [that] figure is the raw, unadjusted, not-tampered-with number." What turned a massive job loss into a respectable gain is something known technically as *seasonal adjustment*. Seasonal adjustment is one of Crudele's precious punching bags.

The truth according to Crudele is to be found in the gray dots in Figure 6-3, which represent the raw monthly tally of jobs in the United States from January 2003 to November 2012. BLS collects payroll data from 150,000 businesses or government agencies every month, selected at random to represent 1,000 industries, 400 geographical areas, and companies of all sizes. Twice a year, in October and February, the statisticians make revisions to bring the CES jobs data in line with the Quarterly Census of Employment and Wages, a more accurate but less regular count of jobs compiled from mandatory state tax records. Such edits have been modest, typically around 0.2 percent, a testimony to the impressive accuracy of the payroll survey, a result of using a massive sample that covers almost one-third of eligible organizations, and of achieving an enviable 80 percent response rate.

The most obvious feature of Figure 6-3 is the sawtooth pattern of the gray dots. The level of employment jerked up and down throughout the decade, with a frequency of about

---

FIGURE 6-3 The Truth According to Crudele: Gray dots show the unadjusted monthly employment level while the black line is the seasonally adjusted, thus smoothed, data.

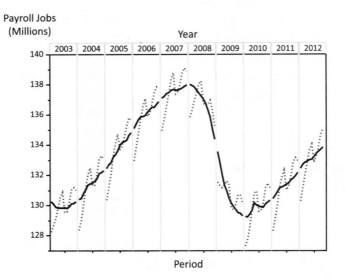

two teeth per year. Call this pattern "Small Teeth." There also lurks "Big Teeth," indicated by the black curve: U.S. employment level steadily climbed from 2003 to 2007, then nosedived, but by 2010, the jobs market had started to mend. This low-frequency sawtooth traces the economic cycle of the decade. "Big Teeth" is also (loosely) known as the *trend line*. Its official name is the *seasonally adjusted data*.

In Crudele's world, the gray dotted line is the truth, whereas the black smooth line is not the whole truth. BLS has exactly the opposite attitude: Having meticulously compiled the payroll data, the statisticians point the public to the black line. Why does the Department of Labor turn gray into black? What information do they sacrifice in the process? Crudele's skepticism can be put into symbols:

GRAY LINE – BLACK LINE =   ?   .

Start with Figure 6-3. For each month, measure the vertical gaps between the trend line and the raw data. Replace the two data series with a single series of their monthly differences. For widescreen viewing, rearrange the side-by-side plots into a grid format. What you get is Figure 6-4. This is a chart that should perk up your NUMBERSENSE. While, in Figure 6-3, the gaps between the black and gray lines appear wildly inconsistent, they are revealed here to follow a stable, seasonal pattern. The yearly curves look nearly identical: Each line rises from a depth of 2 million to a peak of 1 million in the first six months, plunges to negative territory in July, then reverses course, reaching a plateau of just below 1 million in the last quarter of the year.

We just exposed the handiwork of econometricians. They are mimicking the rhythm of the jobs market over a 12-month cycle. Regardless of the state of the economy, employment rises and falls in a predictable pattern, known as the *Seasonal Adjustment Factor*, or simply "Seasonality." Two-thirds of the U.S. Gross Domestic Product is consumer spending, and retailers make half of their annual profits, and 30 percent

FIGURE 6-4 Seasonality: The level of seasonal adjustment ranges widely from month to month, but is quite consistent from year to year.

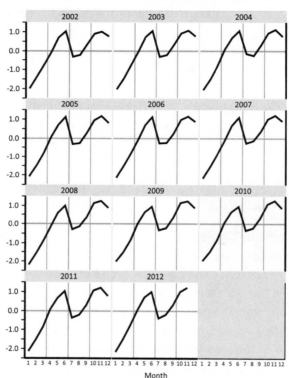

Unadjusted Monthly Change in Payroll Jobs (Millions)

of their annual sales, between Black Friday and the end of the year. The day after Thanksgiving Day is when traditional retailers see their ledgers tip from red to black. The swell of shoppers in the winter months produces a wave of new jobs, many of which are temporary positions soon to wash away in the spring showers. The curves in Figure 6-4 capture this type of seasonal variation.

Completing Crudele's equation yields:

GRAY LINE – BLACK LINE = SEASONALITY

In plain English:

RAW DATA – SEASONALLY ADJUSTED DATA = SEASONALITY

The seasonally adjusted payroll count gets a fair amount of flak, probably because people interpret them as estimates of the monthly employment level. Crudele's allergic reaction is actually quite common. Seasonal adjustment is a big, fat lie, according to its skeptics. They taunt: In January 2012, where should you have sent your resume to land one of those 234,000 newly created jobs? The answer, as you've guessed, is nowhere. Statisticians would not be ashamed to say so. The seasonally adjusted number represents a *run rate* of the employment level. It's the employment level in the average month of the year. Of course, the average month is an invented concept, just like any other statistical average. We know, from Figure 6-4, Januaries lag far behind the average in job creation. If asked how many actual jobs were gained or lost, statisticians would nullify the seasonal adjustment:

RAW DATA = SEASONALLY ADJUSTED DATA + SEASONALITY

They would agree with Crudele that the nation lost 2.7 million jobs in the first 31 days of 2012.

Why does BLS play with the data? Consider Crudele's version of "the truth": the slashing of 2.7 million jobs in January 2012. Is this an ominous sign of an imploding jobs market? Or is it merely a rite of passage, marking the New Year's arrival? You may want to change your answer if you learn that in 2007, when the unemployment rate sank to a cyclical low, 2.8 million jobs vanished in January! Reporting the raw data that depicted a dramatic dive in employment from December to January would be meaningless at best, misleading at worst. The real issue is whether the decline was unusually large or unusually small.

What data analysts have to contend with are two factors fighting for pole position. (We discuss a similar problem in Chapter 8, in the context of fantasy football.) Each month's employment level is impacted by the current state of the economy as well as the month of the year—"Big Teeth" and "Small Teeth." If we let "Small Teeth" go to the fore, we state the obvious. Payroll contracts by nearly 3 million every January. So what? Politicians cannot make seasonality go away. There is no Christmas in January. They believe, however, that the government can alter the course of the general economy through monetary or fiscal actions. So, the econometricians mow down "Small Teeth" in order to see "Big Teeth," using the same equation, now morphed into:

SEASONALLY ADJUSTED DATA = RAW DATA – SEASONALITY

In deriving the Seasonality, BLS staff analyze five years of historical data to establish the average monthly levels. They also do dirty work to make all months comparable. Some of the annoying facts of life, for econometricians, include:

- Differing number of days per month
- Differing number of weekdays per month
- Differing number of workdays included in a paycheck
- The floating nature of Good Fridays and Labor Days

Any of these pesky details could throw off month-on-month comparisons. Forecasters ignore them at their own risk.

Crudele made an example out of an imaginary company that laid off 200 employees after threatening 300 pink slips. He contended that after seasonal adjustment, the report would, to his amusement, trumpet 100 new hires. This would be true if this company had in place a Jack Welch–style rank-and-yank system in which 300 low performers are fired each and every year. In this scenario, statisticians would conclude that 100 people unexpectedly retained their jobs. Compared to prior years, the employment level had indeed improved. The joke was on those who discerned a worsening trend.

Seasonally adjusted data is meant for comparing across months. Explaining the gap between any two gray dots in Figure 6-3 is confusing as well as inconclusive, but comparing two points on the black trend line requires no effort. Reporting the raw counts leads us to a dead end. They tell little about whether the job market is healing or hurting. When the Labor Department announced a seasonally adjusted gain of 234,000 jobs in January 2012, even though in reality millions of jobs vanished, the statisticians saw this loss as being consistent with a slowly mending job market. To season or not to season? Not adjusting is the bigger lie.

3. This Fish Is Spoilt

Like many a disgruntled market observer, the *New York Post's* John Crudele demands "the truth," understood to be "raw, unadjusted, not-tampered-with" data. That standard has a utopian ring to it. It views the human species as destroyer of the earth's purity. What Mother Nature provides us, we will never better. Such a philosophy has recently carved a place in the food business, where the field of data analysis has borrowed much vocabulary ("raw" data, "cooking" the numbers, or "slicing and dicing"). In fine dining, the hot trend is farm-to-table. Some restaurateurs even require clients to eat with their bare fingers or dine in the dark. Some diners gravitate to the no-additives, no-flavor(ing) school of cooking. In animal husbandry, no-hormones, no-antibiotics principles are in vogue. The breast-feeding movement has similar roots. Soon, we might be too ashamed to potty-train our pets. Wouldn't it make sense to leave bodily functions to their natural state?

In data science, the do-no-harm movement goes by motley names, such as "non-parametric," "exact," "distribution free," and "no assumptions." The central idea is the supremacy of making fewer assumptions. Unfortunately, the benefits of these methods have often been oversold. I value them as contributing complementary viewpoints of the data, rather than substitutes. The cost of fewer assumptions is left unspoken; paradoxically, the "exact" analyst can say fewer things, and say so with less confidence. Imagine two tour groups on a safari, racing to sight a leopard before dawn. One tour guide, Mr. Modell, carries a soft torchlight to aid vision and locate animal tracks. The other tour group, led by Mr. Exe, spurns artificial lighting as it disturbs the natural habitat; he instead relies on hearing and smelling. It is true that Mr. Modell's action may have altered nature, and it is also true that his tour group will have more talking points after the excursion. Moreover, Mr. Modell's customers can report that they saw a leopard with their own eyes while Mr. Exe's can describe only what they heard, inferring that the sounds came from the spotted big cat. Of course, both tour guides have their ardent fans. One is not definitively superior to the other. This is a useful metaphor for thinking about statistical assumptions. The trade-off is between seeing things that don't exist and not seeing things that are present. Making fewer assumptions is both a conservative strategy and a cop-out.

Let us also explode the myth of the "raw, unadjusted, and not-tampered-with." All survey data we ever come across have been cooked in one way or another. Consider these scenarios:

1. Students at U.S. colleges grade the courses they took in the past semester. They give ratings to a range of statements, such as "The instructor knows the materials well," from 1 ("strongly agree") to 7 ("strongly disagree"). The last item of the questionnaire is open-ended, permitting students to give any other comments on the course. When the data analyst enters the raw data into a computer program, she notices that about 10 percent of the students may have misunderstood the meaning of the ratings—they raved about the course in the last question ("The best instructor I've ever had!!") but also checked off a majority of 7 ratings. Should the analyst flip the data to align them with the students' true intention?

2. The Bureau of Labor Statistics oversamples Hispanics in each March's CPS survey in order to ensure a sufficient quantity for drawing statistically reliable conclusions about the specific ethnic group. In practice, this means

that the proportion of Hispanics in the sample is about twice their representation in the U.S. population. When compiling statistics about the overall population, should BLS re-weight the survey data to reflect the true relative size of each ethnic group?

3. About 150,000 businesses participate in the payroll survey each month. These businesses are selected at random from a roster of all known establishments in the United States. Despite meticulous planning, some new businesses will be formed after the sample is picked. New entities usually do not respond to surveys until they have hired an accountant. Besides, some businesses collapse after the sample is set, and then there is no one available to fill out surveys. Therefore, the CES sample underrepresents young firms while overrepresenting dying (and dead) companies. Should the government adjust the data to correct the imbalance?

No reasonable person can say no to any of the above. Not adjusting the raw data is to knowingly publish bad information. It is analogous to a restaurant's chef knowingly sending out spoilt fish. The world of Big Data demands more assumptions and fewer bad assumptions.

4. Good Old Washington Statistics

You've experienced that moment. Something mundane managed to make you pause and think. Something unexciting, like the unemployment rate. It's the number that tickles news anchors once every few weeks. It may upset Jim Cramer so much that he unleashes flying objects on his CNBC set. The tantrums last but a day, and then clear. You never pay much attention to that number until this moment. Now, you start to doubt. Really? You connect it to what happened at work. What happened was untoward: cardboard boxes invaded, certain coworkers received a visit, and said coworkers hurriedly exited the building for the last time.

You run through the people you went to college with and you are startled that so many of them have recently lost their jobs, or said they are testing the job market. Several classmates just seem to have vanished: Tom, for instance, whom you call to fix any phone or cable issues, no longer answers your calls. Your best friend, Amy, quit her job so they couldn't fire her, so she said. Your neighbor's son, Steven, moved back after college, and still doesn't have a job. Based on your social circle, you estimate that 20 percent or more must be unemployed. Yet, the Bureau of Labor Statistics (BLS) reported that unemployment never poked above 10 percent, not even at the bottom of the Great Recession. BLS's word is official, and has been since the 1940s. You presume it's just dirty old Washington politics. At this point, your daily routine intrudes and you let the pensive moment slip away.

Many Americans share the same doubt, particularly during the election year of 2012, when pundits expected the economy—more accurately, the unrelenting malaise—to influence a majority of voters. The unemployment rate was more watched and debated than ever before. In the first Presidential debate, held at the University of Denver, Mitt Romney stirred up conspiracy theorists by warning: "Mr. President, you're entitled to your own airplane and to your own house, but not to your own facts." This theme of data manipulation was amplified by Romney supporter Jack Welch, the legendary former CEO of General Electric, who sent a controversial tweet to his 1.4 million followers: "Unbelievable job numbers.... these Chicago guys would do anything... can't debate so change numbers." Curiously, Welch built his formidable fortune while handing out pink slips by the thousands.

Welch reacted a mere five minutes after BLS released the employment report on the first Friday of October, two days after the public slammed Obama's performance in Denver. The unemployment rate for September came in at 7.8 percent, which was 0.3 percent lower than the prior month's; the last time the rate dipped below 8 percent was January

2009, almost four years ago. Welch's off-the-cuff remark, an unsubstantiated accusation, was deservedly and roundly condemned but who among us hasn't harbored qualms about that official statistic?

The amount of 7.8 out of 100 is 78 out of 1,000. You'd think that for every 1,000 Americans, 78 were unemployed during September, which means they did not work even for a single hour during the survey week of September 9 to 15. But this sensible interpretation is way off the standard used by the economics profession. Economists do not consider every American fit for work. Only those who comprise the *labor force* can be employed or unemployed. In fact, there is a complicated set of rules that determines one's employment status, described in a few technical and text documents. I summarize them visually in Figure 6-5.

FIGURE 6-5 Official Unemployment Rate, Sometimes Known as U-3: The number of unemployed persons divided by the civilian labor force. *Marginally attached* and *discouraged workers* are not counted as unemployed. *Involuntary part-timers* are counted as employed. (Not drawn to scale.)

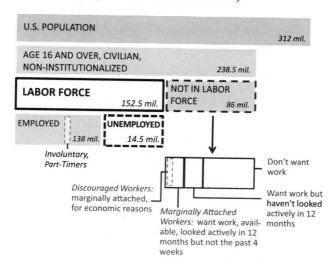

Usually, we think that losing a job takes one's status from employed to unemployed. This is not a given according to BLS rules of counting. Some workers shift from employed directly to "not in labor force," which means they no longer factor into the official unemployment rate. This outflow is the primary reason why the official statistic seems to understate the severity of the jobs recession.

Tom had been a fixer of jammed pay telephone booths for 20 years, and his job was eliminated as these machines disappeared from the street corners. His profession has become obsolete. He is now a middle-aged worker without specialized skills. Hoping to start a new career, Tom enrolled in a nursing program at the community college. Is Tom currently unemployed? One would think so.

Amy just told her manager she's not returning after maternity leave. She worked as an editor for a Manhattan-based publishing house. Though she loved her work, it was not a necessity, as her husband is a star trader at a hedge fund in Connecticut. The couple is raising four kids. She's ready to become a stay-at-home mother. Is Amy unemployed? Probably, as she no longer holds a paid job.

Steven graduated from a liberal arts college in Washington 15 months ago, with a degree in philosophy. At first, he took his job search very seriously. He scoured the job websites, submitting hundreds of resumes. Interviews were hard to come by, and when they did, he faced unfair competition from people with graduate degrees, people with five years of direct experience, and friends of the management. About six weeks ago, he took a break from the job search, exhausted and discouraged. When his bank account dried up, Steven's lovely parents tidied up his childhood bedroom, and invited him home. Is Steven unemployed? For sure.

In the official statistics, neither Tom, nor Amy, nor Steven are considered unemployed—they are excluded from the labor force altogether. Tom will be unavailable for work until he receives his nursing qualification, and starts looking for a job.

Amy doesn't want a job at this time, and isn't looking for one. When Steven graduated, he entered the work force as an unemployed person, despite never holding a job. Five weeks after he suspended the job search, BLS reclassifies Steven from *unemployed worker* to *discouraged worker*—someone who wants a job, is available for work, has actively sought employment in the past year, but has given up in the last four weeks due to economic reasons. Not part of the labor force, Steven's status plays no part in the unemployment rate. (See Figure 6-5 again for details on the several types of not-in-labor-force people.)

How many Toms, Amys, and Stevens are there? Figure 6-6 reveals a marked expansion of this population since the Great Recession hit at the end of 2007. By December 2012, almost 90 million American adults are excluded from the official unemployment statistic (also known as U-3). BLS considers over 36 percent of the civilian, non-institutionalized population as not employable!

How hard can it be to count unemployment? You'd think anyone with basic arithmetic should handle it. Are you out of a job if you don't have one? But what if you don't want one? Did you lose your job if you aren't seeking a new one? What

---

FIGURE 6-6 Growth in the Population Considered Not in Labor Force: Given as a proportion of civilian, non-institutionalized population. [*Source*: FRED, Federal Reserve Bank of St. Louis]

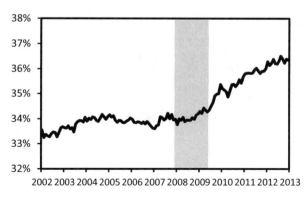

if you decide to do some traveling? What if you do unpaid community service? If you want a job but aren't proactive about it, are you unemployed? If you spend your week reading self-help guides without actually applying to a single job, are you really seeking employment? What if you are attending enrichment courses? Counting, it turns out, is not as easy as it seems. It's certain that different people will hold different views on what counts and what doesn't.

Recognizing the diversity of viewpoints, former Commissioner of Labor Statistics Julius Shiskin developed a broad set of unemployment rates in the 1970s, which evolved into the six metrics BLS publishes today, U-1 to U-6. For example, the U-5 unemployment rate uses a base population that includes marginally attached workers. (See Figure 6-7.) Steven, the recent college graduate who is sick of the job search, would be counted here.

FIGURE 6-7 The U-5 Unemployment Rate: The number of unemployed persons plus *marginally attached workers* divided by the civilian labor force plus marginally attached workers. *Involuntary part-timers* are considered employed. *Discouraged workers* are counted as unemployed. (Not drawn to scale.)

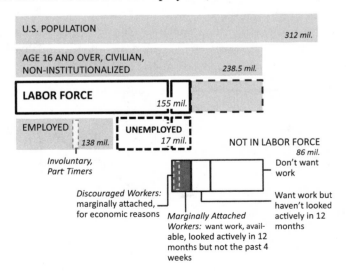

The statisticians at BLS follow a strict set of rules, with roots back to the late 1930s. These are clearly defined, and consistently applied, even if we may not agree with all of them. You are, therefore, unlikely to find experts familiar with BLS procedures who believe the staff has politicized the generation of statistics. That's why Jack Welch's rant was widely ridiculed. That said, you will find many analysts who create their own flavor of the unemployment rate.

People are fickle when answering surveys. In Chapter 1, we saw how surveys can fail to elicit accurate data on the career prospects of law school graduates. What do people really mean when they tell interviewers they don't want a job? Can 80 million people afford not to earn a salary? Say you like to be conservative, and assume everyone wants a job. This most expansive definition of un(der)employment, as shown in Figure 6-8, comes in at a whopping 42 percent.

FIGURE 6-8 Another Unemployment Rate: The number of unemployed persons plus those not in the labor force divided by the civilian, non-institutionalized, above age 16 population. (Not drawn to scale.)

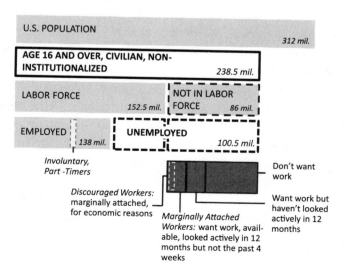

FIGURE 6-9 Employment-Population Ratio (2002–2012): This statistic plunged during the Great Recession and remains stuck at the bottom. [*Source*: FRED, Federal Research Bank of St. Louis]

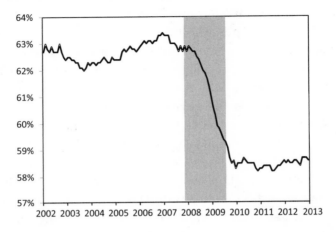

The inverse of the unemployment rate depicted in Figure 6-8 is known as the *employment-population ratio*. Some economists consider it more informative than any of the six BLS metrics. This measure paints a disturbing picture of the nation's employment situation. (See Figure 6-9.) We are stuck at the bottom and unable to get up. When evaluated together with the U-3 metric, we deduce that the drop in the official unemployment rate has more to do with the people "not wanting jobs" than people finding jobs.

Any of these rates can deceive. The employment-population ratio does not distinguish between young college graduates and retirees, so it tends to overstate unemployment. Besides, there are always people like Amy who don't need to work. That's one of the reasons why economists argue the unemployment rate cannot, and should not, be zero. Even this statement is problematic: By now, you should know that the official unemployment rate is not being bloated by people like Amy, who are excluded from the count!

5. Crudele says "Oops"

On August 4, 2012, John Crudele, the *New York Post* financial columnist who, for years, has humored the government econometricians respectfully, threw in the towel. In his newest piece, he ranted:

> It's cheating time in Washington. I've long believed the Labor Department's monthly employment statistics are horribly inaccurate. So bad, in fact, that they are hardly worth compiling. But I never thought the numbers were fudged—until now.

The object of his derision is the so-called *Net Birth/Death Model*, perhaps his favorite punching bag of all time. He has variously labeled this object: "the nearest thing to a fraud perpetrated by Washington," "Class A razzmatazz," "figments of the imagination of the Labor Department's computers," and so on. The Birth/Death Model creates "make-believe jobs that the government can't really prove exist." For example, the Model added 206,000 jobs in May 2011, a month in which jobs growth was estimated to be 54,000 after seasonal adjustment. Twice a year, BLS issues "benchmark" revisions to the CES data after reviewing the more accurate Quarterly Census. In most years since 2000, 100,000 to 200,000 jobs, between 0.1 and 0.2 percent, are added or subtracted from the annual jobs count. Crudele cheekily calls these the "Oops Reports." He complains that the corrections largely reverse the ill-conceived Birth/Death adjustments.

The reality is the opposite. These earlier adjustments bring the statistics closer to the Quarterly Census months in advance of the benchmark revisions. In order to appreciate the Birth/Death adjustment, we must look at a counterfactual analysis. (I defined this concept in Chapter 3.) The question to ask is: How large would the benchmark revisions have been if the numbers were not previously adjusted for net business births and deaths? According to a 2008 study by BLS economists, the subsequent corrections would have been multiples as large.

The Net Birth/Death Model is designed to solve a problem of selection bias, which I listed above as Scenario III. The payroll survey, if unadjusted, would systematically undercount jobs created by start-up firms, and simultaneously overestimate jobs eliminated by business closure. Crudele correctly observes that these jobs are not countable, not by BLS, not by anyone else. The BLS Model makes a guess based on historical data. The scale of these changes is, in fact, dwarfed by the number of jobs in the country—the May 2011 addition was 0.15 percent of 130 million jobs. What if the adjustment is a means to get closer to "the truth," without being attributed to correcting the selection bias? I reckon the audience would be less confused.

Numbersense begins with looking at the data. Keep your hands off, though, until you have dug out the excruciating details of how the data is collected. There is nothing more fundamental in an applied statistician's toolkit. So John Crudele is right on this point. But raw, untampered-with data almost never yields the answer. Some form of adjustments—whether it's seasonality, or bias correction, or others—are like the dressing on top of a salad.

How Much Did You Pay for the Eggs?

Recall when you last shopped for groceries. Do you remember what you purchased, and how much you paid for each item in your shopping cart? If you bought a carton of milk or juice, do you know if the price you paid was over or under the average? When you answered the previous question, what did you mean by the average? Was it the normal price at the specific store, or the median price among several stores in your neighborhood, or something else? Do you remember if the milk was one of the store's weekly specials? Do you remember if you redeemed a clipped coupon? Do you remember if you picked up a new brand of juice because of a promotional offer? Did you switch from Tropicana to Odwalla, or from Minute Maid to SunnyD?

If you are like the average shopper, you'll have difficulty coming up with these answers. When it comes to remembering prices, we are hopeless.

Businesses have long known—and exploited—our price amnesia. In the late 1980s, two marketing professors, Peter Dickson and Alan Sawyer, collaborated with a large supermarket chain to measure just how clueless consumers were about purchases they made only 30 seconds or fewer ago. Researchers intercepted shoppers immediately after they placed certain target products—such as coffee, toothpaste, and margarine—into their shopping carts. Almost everyone consented to answering a few questions when offered a $1 gratuity for participation. To raise the chance of finding price-conscious shoppers, part of the study was conducted in late January, when household budgets were stretched after the winter holidays. Did people know how much the stuff in their shopping carts cost? Were they aware of any special deals that applied to those items? The results of the 800 or so interviews conducted at four different branches of the chain were disturbing.

After arriving at the display, the average shopper moved along within 12 seconds, but the majority could not name the correct price of the item they just took off the shelves. The average error was 15 percent of the real price. One out of five shoppers could not even offer a guess of the price. Their awareness of special discounts was even poorer. This supermarket chain heavily advertised specials in newspapers and on television, using the phrase "Cost Cutter Bonus Buy" accompanied by a scissors symbol. In addition, the management placed bright-yellow labels with the slogan and scissors right next to the standard black-and-white price labels on the store shelves. And yet, three out of five had no clue if the item in their shopping cart was on special or not; estimates of the price reduction given by those who could suffered an average error of 47 percent.

The jaw-dropping findings didn't stop here. The researchers learned that people who shopped frequently for an item were equally as hopeless as the rest. Finally, the professors performed an aided brand awareness test, similar to the one mentioned in Chapter 1, on the hunch that some shoppers could recognize the special price label even if they could not recall the exact price. In yet another surprise, only 54 percent of the participants managed to pick out the correct price label from a choice of three.

This line of research asks deep questions of the foundations of modern economics. In a market economy, prices are supposed to capture all there is to know about supply and demand. Producers and consumers are predicted to respond to these prices. When a half or more of the population are blatantly inattentive to price tags, we wonder if the economic profession has gotten this core assumption wrong. Dickson and Sawyer thought consumers with stronger motivation to consider prices would perform better in their study, but it turned out those who shopped at the inner-city store were even more clueless about the amount they spent on groceries. Marketing experts have long ago abandoned many economic principles that are at odds with reality. Behavioral economists are now tackling this kind of challenge, and their insights may well modernize the foundations of the discipline.

Now, take the side of the store manager. For a gallon of milk, we require a target price of $3.50 over the next four weeks. We can, unimaginatively, set a fixed price of $3.50. But a good portion of our customers love the game of coupons and deals. We can, for example, charge a normal price of $3.60, and one day a month, run an irresistible bargain price of $1.50. Alternatively, we can advertise a weekly special of $3.00 on a regular price of $3.60. All three pricing schemes produce an average price of $3.50. Which strategy would yield the most revenues? The winner depends on how our customers respond to discounting. That in turn depends on how they process the prices. Here are a number of possibilities to consider:

- *Availability*: People take what comes to mind first. Behavioral psychologists Daniel Kahneman and Amos Tversky, whom we met in Chapter 5, are champions of this theory.
- *Recency*: Perception is affected by the most recent price encountered.
- *Frequency*: Customers remember the price that appears most often.

- *Average*: Customers have a mental image of the average price. This suggests that they intuitively sense the average value of a set of numbers.
- *Median*: Customers have a mental image of the median price. This requires that they spontaneously discard extreme values.
- *Extremes*: Perception is swayed by unusually large or small numbers.
- *Losses*: Customers pay undue attention to price increases because they regard price increases as financial losses.
- *Numerosity*: Customers perceive a better deal when savings are divided into numerous small installments rather than applied in total to a single purchase.

There is as yet no definitive research on how consumers perceive prices. It's not even clear that everyone favors the same set of heuristics. The decision criteria may vary by the type of purchase. For durable goods not replaced often, like stoves and ovens, it's irrelevant to talk about frequency, average, median, or numerosity. Big-ticket items and petty spending surely are not given equal consideration. Perhaps Kahneman and Tversky's perspective is the broadest: All the other criteria pinpoint which price becomes "available."

1. There You See It, There You Don't

How much are prices increasing? That's one topic on which there is no shortage of opinions. If in doubt, ask Mom (or whoever carries the wallet in your household). My mother is an exacting shopper with a keen eye for the bargain. I count on her to know which stores to buy what products, what time of year to window-shop and when to open up my wallet, which coupons can be combined, and when to use the percent-off discount versus the dollars-off variant. I asked her about the price of groceries. She noticed paying more for eggs and bakeries. The price of fresh fruits and green vegetables, always plentiful in California, hasn't moved much, especially if she stuck to the specials. Coffee definitely costs considerably more. The politicians were pumping up bridge tolls, she added, and also traffic fines.

For more than 30 years, the *Survey Research Center* at the University of Michigan has routinely asked people: "By about what percent do you expect prices to go up or down on the average, during the next 12 months?" The responses are compiled into the *Inflation Expectation Index*, which the Department of Commerce uses as one of 11 components of its *Index of Leading Economic Indicators*. In the first half of 2008, the median person expected prices to rise by 5 percent annually. But the individual assessments wandered all over the map. Over a quarter of the respondents in July 2008, for example, believed the year-ahead inflation would fall between 10 percent and 20 percent. Around this time, the media was reporting the official inflation rate, called the *Consumer Price Index (CPI)*, of about 2.5 percent.

Researchers are baffled by the diversity in opinion on a topic of such importance and pertinence. Not only does economic theory rely on consumers responding to price changes, but the government also anchors various social spending programs to the inflation rate calculated by the Bureau of Labor Statistics (BLS). Moreover, the Federal Reserve's mandate includes price stability. The great CPI puzzle is why perceived price changes stray so far from the official inflation rate.

How much are prices increasing? I hope I have convinced you that this is no simple task. Which heuristics do you use to estimate the rate of inflation? Which purchases come to mind? Are you thinking about repetitive purchases such as food and toilet paper? Are you taking into account irregular outlays, like television sets or sofas, which have big ticket prices? What about rent and tuition expenses? How confident are you that you remember the prices you paid, let alone how those prices have evolved over time? One piece of the CPI puzzle is our inattention to everyday prices. But that is only one piece, and there are several others.

2. The Discontent of Being Averaged

The human brain is lousy at estimating prices from memory or intuition. But you can grab pencil and paper to work out a personal inflation rate methodically.

Begin with a listing of all out-of-pocket purchases in the past two years. This list includes both goods and services, as well as recurring and one-off spending. Some items are elusive, such as insurance premiums taken directly out of paychecks, other forms of scheduled payments, and stuff bought using gift cards. Deals and discounts make a mess, while returns and price adjustments are a pain to track. Small cash purchases quietly stack up: Two Starbucks a day over a year cost more than a month's average rent ($804).

Now, sort the items by type: food, energy, communications, and so forth. More than likely, there is a consistency in how you split up your spending among these categories from one year to the next. If the distribution of expenses shifted markedly, as can happen with certain life events—such as marriage, childbirth, and relocation—the normal notion of an inflation rate loses any meaning. *Inflation* is usually defined as the increase in cost to maintain a stable quality of life. Someone who wins a big promotion at work may start to live more extravagantly, say by shopping at Whole Foods for pricier, organic produce or by building a vacation home in the Colorado slopes. The consequent growth in household expenditures defies the common sense of inflation as a change in prices paid.

Use the prior year as your reference year. The items bought last year constitute your typical "basket," an example of which is shown in Figure 7-1:

FIGURE 7-1 A Sample Consumer Expenditure Basket

Expenditure Category	Spend $	Spend Weight	
	Reference Year		
Food	9,000	15%	
Housing	18,000	30%	
Apparel	2,000	3%	
Transportation	10,000	17%	
Medical care	6,000	10%	
Entertainment	4,000	7%	
Education	2,000	3%	
Others	9,000	15%	
Total	60,000	100%	

How much did the same basket cost in the following year? For staple goods, such as Wonder Bread and Ben and Jerry's ice cream, because you purchased them in both years, the change in prices is directly obtained. Be careful as manufacturers often disguise price hikes. Take that jar of Skippy Peanut Butter, and feel its bottom. A few years ago, Skippy added a dimple to the base, skimming about 10 percent of the volume. Further legwork arises for similar but not identical items. You may consume six pounds of cookies each year, but the Fig Newtons of last year aren't exactly the Pepperidge Farm Milanos of this year. The bag of Chips Ahoy! isn't the same as chocolate chip cookies from the boutique bakery. Oreos in a vending machine have different price tags from Oreos at Costco. If you didn't buy the identical item this year, you have to figure out the current price for the last-year item.

That sounds painful, until you deal with the cable company (no surprise). It raised its tariff yet again, while adding ten channels to the bundle. Three of those channels are in Spanish, a language you do not speak. One is the Cooking Channel, a derivative of the preexisting Food Channel, only "grittier, edgier, and hipper," adjectives that none of your friends would attach to you when they are stoned. One shows classic movies, which excites you a wee bit. Several are high-definition clones of popular channels. A couple of the high-definition channels you can find over the air. How much of the price increase was due to inflation, and how much was justified by more and better programming? What value does each channel contribute to the bundle?

Luckily, the Bureau of Labor Statistics has done the heavy lifting. They publish hundreds of basic price indices. If 30 percent of your basket consists of restaurant bills, you can look up the "Consumer Price Index for All Urban Consumers: Food Away from Home" for your region, which measures how much the cost of dining out has changed from one year to the next. Your personal CPI is the average of these component indices, weighted by the relative importance of each expenditure category.

The process just outlined describes 90 percent of how the BLS computes the CPI. One key asset the professional data collectors have is a set of rules that resolves the counting challenges, such as new packaging, quality improvement, and discounting. Because the agency issues one number for the entire nation of 104 million urban households, it is no wonder our perceived inflation rates differ from the experience of "the average American." This gap is the second piece of the CPI puzzle.

The average American? As I discussed in *Numbers Rule Your World*, you can travel to each corner of the 50 states, and you won't find one person whose behavior mirrors the average Joe from the pages of the *Statistical Abstract of the United States*. The average is like everyone, but no one is the average. The policymakers in Washington, D.C., though must take actions for the benefit of the whole nation, so they worry about the average rate of inflation. We expect the CPI to reflect our personal experience; it will not, and cannot, simply because the process isn't designed for this purpose.

Let's revisit how inflation is measured. This time, we watch how BLS statisticians add averaging to the mix. It's impossible to audit every household budget in the nation. The CPI arises from a series of surveys. These questionnaires collect data from urban consumers only, covering 80 percent of the population. So, if you live in a rural area, your experience isn't included. The basket is then obtained by weighting and merging the responses to how income is being spent. The food budget of a vegetarian who mostly eats at home looks nothing like that of a meat-lover on the Atkins diet who never cooks. When the answers are tossed together, they become parts of the average American, who consumes a little bit of everything. Similarly, most people either rent or own a home, but the average American does both in proportion.

Now, look inside the basket. The BLS places items into 200 odd groups: Eggs is one such group. The actual price paid for eggs is not stable. It depends on the size of eggs, the quality of eggs, where you are located, where you shop, and even uncontrollable factors like coupons, weather, cost of fuel, and so on. Through surveys, the BLS determines what type of eggs the average American buys, and from what types of outlets. Each month, field staff visits a sample of stores to collect price quotes. In the case of eggs, they gather 10 to 15 quotes in each major city, and about 5 quotes per smaller city. A price quote is something that looks like this:

ONE DOZEN AA EGGS, LUCERNE BRAND, SOLD AT
SAFEWAY (BERRYESSA ROAD, SAN JOSE, CA), $2.49

At each store, the data collector selects a subset of all egg items on sale, according to their popularity. The prices are then averaged. Every month or two, the selection of items shifts; every three months, the choice of outlets is rotated.

How would this average price compare to what you pay for eggs? If you shop at a farmer's market, you pay a different price. If you eat exclusively Trader Joe's cage-free eggs, you pay a different price. If you prefer eggs by the half dozen, your unit price is higher. If you live in the Midwest, your price is lower. If you have an egg allergy, you pay a heavy price. The single number from the BLS is never going to match everyone's experience.

All in all, the published CPI number is a roll-up of a ton of painstaking details. Each basket contains over 200 categories of items. There are in fact baskets for each of 38 regions. More than 8,000 basic indices are created for each combination of region and expenditure category. From these, the BLS produces regional indices, item category indices, and various aggregate indices.

We are deep in the world of Big Data. Anyone can retrieve the thousands of indices. Policymakers should be crafting smarter economic policies that reflect the diverse patterns of consumer spending. There is no excuse for one-size-fits-all policies. Everyone can build a personal inflation rate. Do not expect this rate to match the official CPI, which is a statistical average. Instead of worrying about the average value itself, we ought to focus on the variance around the average. This gap is, in fact, quite informative—it represents how our spending habit differs from that of the average American.

3. Whose Core?

The first piece of the CPI puzzle is our cluelessness about how much money we forked over for anything. Even if we could precisely recall prices, one aggregate number does not capture millions of individual experiences: The discontent of being averaged is the second piece of the puzzle. In case we defeated the statistical gods to make ourselves into John or Jane Average, our computed rate of inflation would still not match the official statistic. And this is when we realize we don't inhabit the same world as the economists who advise the government.

Since the 1970s, the high priests of economics have sold U.S. policymakers on something they label the "core" inflation rate to differentiate it from the number we've been talking about in the last sections, which the economists call the "headline" CPI, as if it's fit for newspaper columns, and not for serious people. The *core inflation rate* is the CPI of all expenditures except those of food and energy. (The Bureau of Labor Statistics, which first published this supplementary data series in 1977, never uses the term "core," preferring "all items CPI less food and energy.")

Core has several meanings:

- A central and often foundational part of
- A basic, essential, or enduring part of
- The essential meaning of
- The inmost or most intimate part of

Economists apparently use the meaning of "foundational" or "essential" when they use the noun-adjective "core." They assert that the core inflation rate more accurately measures the long-term trend in general prices in the country. The wild swings that occasionally hit food and energy prices are but a distraction. The NUMBERSENSE attitude to such a pronouncement is to check the evidence, and not accept it at face value.

In Figure 7-2, we see the striking effect of ignoring food and energy spending—a type of statistical adjustment known as *filtering*. The chart is a tale of two lines. Between January 2007 and October 2012, according to the "core" inflation line, the U.S. economy was gliding along nicely, at least as far as the prices of goods and services were concerned: The inflation rate was stable, trapped in a narrow range between 1 percent and 3 percent annually; this means that the rate at which prices were changing was steady, not that prices weren't rising. If, at the risk of angering your economist friends, you steal a peek at the "headline" inflation line instead, you would probably think that you and they live in different universes! What you see in the "headline" or all-items number is prices climbing at an annualized rate of 4 percent or 5 percent in the early months of 2008, then a scary breakdown for about a year through the middle of 2009 around which time the general cost of living actually eased, and a subsequent, slow, stuttering reversal. These ups and downs vanished from the "core" inflation line.

Ask yourself what happened to the U.S. economy in the five odd years since 2007. Or, look back at Figure 6-3, which shows the employment situation during this period. Which of the two stories echo the real economy?

Consider a pilot who safely landed an airplane after a particularly tumultuous flight across the continent. His flight crew congratulated the passengers on the uneventful arrival, just as they do every day. For the customers, the enduring memories of the journey were the patches of turbulence, the violent thrusts, grasping on to the seat partitions, holding

---

FIGURE 7-2 "Core" (---Line) versus "Headline" (Solid Line) Inflation Rates: Year-on-year unadjusted rate

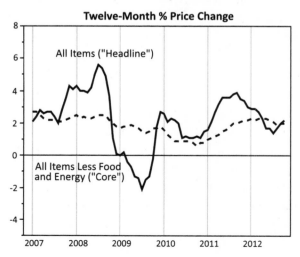

hands with their loved ones, and balancing their half-filled beverage cups. Therein lies the difference in perspective, which is the third piece of the CPI puzzle. The economist thinks like the pilot; we feel for the passengers.

The list in Figure 7-3 contains some of the major categories included in the CPI. I have separated them into two groups. See if you can figure out why some items are placed on the left, and others on the right. Think about where, when, and how you pay for these goods or services.

Can you recognize the difference in shopping patterns? The items in Group A are purchased very frequently. There is barely a week that goes by without having to pay for food or fuel. If you don't cook, you'd eat out. Everyone is constantly powering gadgets. By contrast, you only buy items in Group B once in a while. Once we move into an apartment, the rent is fixed, and you are not revisiting that decision for a while. Vehicles or houses are only purchased a few times in a lifetime. So, from one day to the next, your thoughts center on items in Group A. If you are asked about prices, food and gas prices naturally come to the fore. This is the concept of *availability*. Unlike economists, most people regard food and energy to be "core" expenses. They are certainly "basic" and "essential" to

---

FIGURE 7-3 Major Categories of Consumer Expenditures

Group A	Group B
MEAT, POULTRY, FISH, EGGS	RENT, MORTGAGE EXPENSES
CEREALS, BAKERY PRODUCTS	VEHICLE PURCHASES
FRUITS, VEGETABLES	HEALTH INSURANCE
DAIRY PRODUCTS	APPAREL
FOOD AWAY FROM HOME	TUITION
ELECTRICITY	
FUEL OIL, OTHER FUELS	
GASOLINE, MOTOR OIL	

our very existence. On average, they account for one in four dollars we spend during the year.

Nonetheless, economic advisors have convinced government officials that the items in Group A are meaningless. The advisors are not just arguing that those things are less vital, they literally assign an *importance rating* of zero to them. Here is another way to understand Groups A and B. Items in Group B are the major expenses that largely determine the "core" inflation rate while Group A contains items that don't factor in "core" inflation. Such filtering has a predictable statistical result of loosening the correlation between "core" CPI and food or energy prices. This misalignment can be stated less technically: The official inflation number contradicts the daily experience of consumers.

4. Drill, Baby, Drill!

The next time an economist insists that food and energy prices are too noisy and thus useless, you should show him or her Figure 7-4, especially the left part of the chart, and watch the economist squirm.

You are conditioned to expect the food CPI to jerk up and down in some unpredictable manner, but you won't see it in the recent data. In fact, food prices have moved in lockstep with the all-items CPI. Lately, the feared gyrations are confined to energy prices. Scientists who investigated this trend learned that variability has been tamed because we now eat more processed foods and we dine out more often, where menu prices are sticky. This trend is highly meaningful, and food experts expect it to stay. It is also easily spotted, unless one cares only about "core" inflation, in which case food prices and their stabilization have landed in the same dumpster. When explaining the concept of "core" inflation, many economists regurgitate something that hasn't been true for a while.

Now we get why statisticians hate throwing data away. They do sometimes discard bad data, but high variability is

FIGURE 7-4 Food and Energy Component CPI: Food and Energy Indices Relative to All-Items CPI

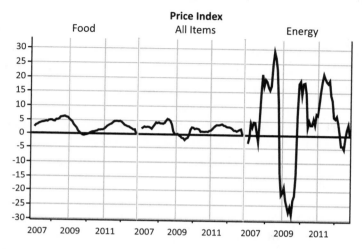

a poor proxy for badness. Place yourself as the manager of quality at a factory of hand-made leather shoes. Color varying from one pair to the next is perfectly acceptable, and may even be considered a feature of top-quality cowhide. But you will surely reject the shoes damaged by scratches from a sharp instrument. When the Bureau of Labor Statistics insists on describing "core" inflation as "All Items Less Food and Energy," they are also making a professional stand against unwarranted cleansing of data.

The economists are wise to notice that certain prices swing more than others. Their mistake is to shove this meaningful feature out of sight. Statisticians typically apply a strategy of *disaggregation*: They disassemble the data, and drill into the components individually. In *Numbers Rule Your World*, I showed how designers of the SAT test and insurers use this general principle. Inflation statistics follow in much the same way.

The BLS makes available *thousands* of disaggregated price indices as part of its monthly CPI release, even though the national media only ever talks about "core" inflation or "headline" inflation. The BLS offers component indices focused on food or energy or any of scores of expenditure categories. They have inflation rates for eggs and furniture and cable television subscriptions, and pretty much anything you can imagine. There are regional indices covering different parts of the country. There is even an experimental index for older Americans, adapted to their distinct spending pattern.

The BLS invites us to drill, baby, drill! I took this opportunity to validate some of Mom's observations on the direction of grocery prices. Not all food groups are born equal: The flatness of the aggregated food CPI line masks the incredible diversity in its constituents. (See Figures 7-4, 7-5, 7-6, and 7-7.) On eggs and milk, the BLS reported an eyebrow-turning 20 percent leap in prices from mid-2009 to mid-2011, confirming Mom's shopper credentials. What is surprising, though, is that by late 2012, stores were selling eggs and milk at roughly the same prices as they did at the start of 2008, as the recent price hike essentially reversed the acute slide during the Great Recession.

What about fruits and vegetables? Living in the Golden State, Mom loves shopping for these items, as supply is generous, quality is superb, and prices are extremely reasonable. The data tells us this situation isn't unique to California. Nationally, the prices of fresh fruits and vegetables have held their ground in spite of the economic upheaval. By 2012, the average price level has more or less returned to the level of early 2008. By contrast, processed fruits and vegetables have become 20 percent more expensive since 2008.

The coffee CPI also supports every coffee drinker's lament. The price of coffee has indeed suffered runaway inflation: In little more than 12 months, our caffeine addiction cost 25 percent more.

Perhaps the survey respondents who imagined inflation was running at 10 or 20 percent weren't really out of touch. They may have come to this judgment based on some of their most frequent purchases—like coffee, milk, and eggs. They seemed to have neglected other categories in which prices have dropped, such as clothing and home furnishings.

The CPI statistic is layered like a set of Russian Matryoshka nesting dolls. Taking apart the all-items number, we find the *Food Index*, from which we can split 60/40 into the *food at home* index and the *food away from home* index. The food at home index is split again into (in order of importance):

- Meats, poultry, and fish
- Fruits and vegetables
- Cereal and bakery products
- Non-alcoholic beverages
- Dairy and related products
- Sugar and sweets
- Fats and oils
- Eggs
- Others

FIGURE 7-5 How Prices of Selected Foods Changed Since 2008—Eggs and Milk: Eggs and milk prices followed a similar trajectory, first falling almost 20 percent by mid-2009, and then returning to the 2008 level by late 2012.

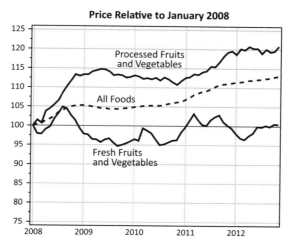

Price Relative to January 2008

FIGURE 7-6 How Prices of Selected Foods Changed Since 2008—
Fruits and Vegetables: Processed fruits and vegetables became
20 percent more expensive while the prices of fresh fruits and
vegetables remained relatively stable.

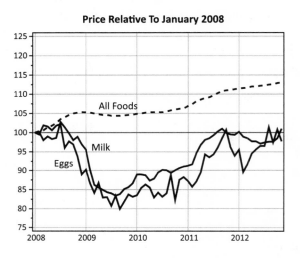

FIGURE 7-7 How Prices of Selected Foods Changed Since 2008—
Coffee and Bakery Goods: Coffee prices rose by 25 percent
between mid-2010 and late 2012 but have since moderated;
inflation of bakery goods prices was more severe in the first half
of 2008 but has since followed the general food index.

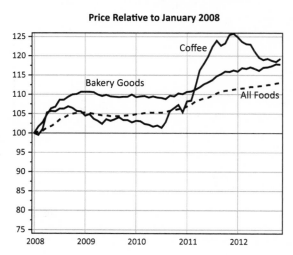

The design of each inner figure must vary, or else the fun of Russian dolls oozes away. Too much emphasis on aggregates in statistics has the same effect as gluing shut one of the middle dolls, justifying this unkind act by arguing that the smaller ones have an identical look. In the case of the "core" CPI, a couple of inner dolls are taken out of the set while collectors are told the value remains the same because those figures are "ugly."

5. Awe of the Average

Economics reporting suffers from an awe of the average. The Consumer Price Index represents the experience of no one. It is the average change in prices at an average retailer for an average set of items selected to represent specific item groups in an average basket of goods and services bought by the average consumer in an average region of the country. We keep hearing about this one number, and often are bewildered as to why the official statistic jars with our personal experiences as consumers. If the reporters talk about a different number, it is inevitably the "core" inflation rate. This metric excludes purchases of foods and energy, things that for the most part shape our perception of price changes.

Journalists on the economics beat have yet to wake up to Big Data. The Bureau of Labor Statistics makes public thousands of price indices covering geographic regions, expenditure groups, and various definitions of inflation rates, and yet we seldom hear about them in the news. Disaggregation unwinds the averaging process, and the component indices tend to make more sense to us. When data is plentiful, we should appreciate the diversity of its components. Two strategies that sometimes backfire are averaging and filtering. The former stamps out the variety while the latter casts dark shadows.

PART 4

SPORTING DATA

Are You a Better Coach or Manager?

One of my favorite neighborhood trattorias in New York City shuttered recently. Being there, in the dining room with yellow stuccoed walls, wooden accents, and rustic linens used to bring back memories of pleasant dinners in a Tuscan farmhouse. Rooted at the chef's bar, I loved watching cheeses, hams, olives, and breads being assembled on plates, and peering into the brick oven in the back corner, where the staff prepared roast pork, octopuses, and peppers. Bellavitae was a mid-priced Italian restaurant nested at an unlikely alley in a city straddling a grid of streets and avenues; Minetta Lane had but one other commercial outfit, an independent theater.

I could imagine the chef's dismay at the lukewarm review by Frank Bruni, then the influential food critic at the *New York Times*. "Much of the menu at Bellavitae is devoted to food that requires plating more than it really does cooking," he began. He later elaborated, "The more actual cooking was involved in a dish, the less successful that dish was likely to be." I could feel the sting of those remarks every time I munched on the crostini with chicken liver pate at the bar since.

Bruni's thorny critique popped up in my head one day during a conversation with my friend, Jay, on the unrelated topic of fantasy football. Jay is a freelance photojournalist, having cut off a career in publishing during which he compiled several statistics titles, among other textbooks. He spent his college years in St. Louis, Missouri, and still roots for the Rams football team although for over 10 years, he has lived in Boston, San Francisco, and Hong Kong.

In 2006, Jay joined the Tiffany Victoria Memorial Fantasy Football League (FFL), naming his team "Tuff Toes." Having placed well in small-time, non-monetary leagues, he was eager to test his mettle "in the bigs." Fantasy football has swept over the United States since the mid-1990s when CBS, among others, launched websites hosting fantasy leagues, providing fans with easy access to schedules, statistics, scores, and tools. In 2011, an Ipsos poll found 24 million FFL participants, 20 percent of them women.

Halfway through the 2011–2012 National Football League (NFL) season, Jay was dissecting the data to assess his own strengths and weaknesses. Particularly, he wanted to know if he should spend more time optimizing his roster of players—through wheeling and dealing—or selecting game-day squads from the roster.

Jay was inspired by a legendary comment by Bill Parcells, the legendary NFL coach who earned his reputation in the 1980s by turning the New York Giants from perennial under-achievers into double Super Bowl champions (1986, 1990). In 1996, Parcells, who had moved to the New England Patriots, was locked in a power struggle with Robert Kraft, the team's owner. The unhappy coach famously uttered, "If they want you to cook the dinner, at least they ought to let you shop for some of the groceries." His clever analogy depicts the delicate relationship between the general manager and the head coach of a football team. Kraft favors the traditional division of responsibility:

- The *general manager* constructs the roster of players by drafting, trading, and using the waiver wire, while keeping an eye on the salary cap.
- The *coach* selects game-day squads, designs the strategic approach for each opponent, and makes tactical choices on the field.

At the time, Parcells's coaching ability was beyond doubt. The coach, nevertheless, wasn't satisfied with the roster he had to work with. When Kraft refused to wrest the managerial duties from his long-time personnel director, Parcells bolted to the New York Jets.

For the novice, fantasy football is simple, as long as you don't think of it as football. Instead it resembles an investment game, in which players compete to assemble the most profitable portfolio of stocks within a fixed number of weeks. The "stocks" are players in the National Football League, and the "stock prices" are computed at the end of each weekend's games by a scoring formula of your league. Your "portfolio" consists of nine players you activate each week from a roster of 14 players. The five bench players do not earn any points, as when you insert interesting stocks in a watch list. The scoring formula is a combination of various real-life statistics. For example,

- A quarterback (QB) earns points by throwing over 400 yards.
- A wide receiver (WR) earns points by amassing over 100 yards.
- A kicker (K) earns points by scoring four field goals.

Each skill position has its own set of metrics. In essence, you place bets on which players will impress in the following week's matches. You are also wagering on injuries since an inactive player in real life earns zero points in fantasyland. Like bargain hunters who scour the Sunday newspaper inserts, fantasy football fans monitor injury reports for every snippet of data that might ignite their imagination.

There are a couple of twists. The selected squad of nine must comprise of:

- One coach (C)
- One defensive and special teams unit (D/ST)
- Seven offensive players—one quarterback (QB), two wide receivers (WR), one tight end (TE), one running back (RB), one kicker (K), and one wild card (usually a second QB, a second RB, or a third WR)

Think of skill positions as "asset classes," such as health care, utilities, and high-tech.

You can only activate players you own that week. The roster is set before the first week of the season by conducting a *draft*, in which the team owners pick players in some prescribed order. Over the course of the season, the roster is edited through wheeling and dealing with other fantasy teams. For each week, teams are matched up in pairs, and the goal is to score more points than your opponent. (These rules may vary slightly between different fantasy leagues.)

In the description above, no distinction is made between owners, managers, and coaches because FFL owners manage as well as coach their teams, exactly as Bill Parcells desired. While fantasy coaches are stripped of strategic and tactical decisions, many of them get raving mad at real-life head coaches for deploying tactics that harm their weekly points total, such as not running up the score in a blowout or not giving their star player enough touches of the ball.

Jay's quarterback, Drew Brees of the New Orleans Saints, started the 2011 season with a bang, gaining 419 yards and three touchdowns at Green Bay. Brees's heroics in Week 1 produced 34 fantasy points, which gave Jay's team a 20-point advantage at the QB position against his opponent who activated Eli Manning, of the New York Giants. Little did it matter that the Saints lost their season opener to Green Bay, notwithstanding the shiny numbers by their stud QB. After 13 weeks of competition, Jay's team, Tuff Toes, amassed 1,297 points, tied for second of 14 teams in the Tiffany Victoria Memorial FFL. However, his head-to-head record of five wins and eight losses disappointed, placing Tuff Toes three slots from the bottom (tied with two other teams). Jay wanted to improve his finish next season, but the conflicting outcomes were a puzzle. Should he listen to Parcells and spend more time shopping? Or heed Frank Bruni's assessment that a proper chef must accomplish more than plating fresh ingredients? Jay showed me some preliminary analysis he did; I took that and expanded its scope.

1. Inviting a Statistician into your Kitchen

Jay's conundrum has the scent of a classical statistical problem. We wish to explain a pair of related outcomes, namely, the *fantasy points total*, and the *win-loss record*. These metrics varied widely among the 14 teams. (See Figure 8-1.) The points total ranged from 988 to 1,380; and the number of wins, from three to 10. What are the factors contributing to such variability? Following Bill Parcells, we consider two key factors, *managerial acumen* and *coaching ability*. This feels right but feeling right isn't the same as NUMBERSENSE. The proposal needs validating. Saying so does not make it so. Does the simple two-factor model explain what actually happened? It is possible that only one of the two factors matters. It may be that the two factors combined still don't provide the full picture. There is, also, the luck of the draw.

The world is filled with such problems. In *Numbers Rule Your World*, I describe how psychometricians explain differential results in standardized tests among different groups of students by separating the effect of ability from the effect of test item bias, properly known as *differential item functioning* (*DIF*). Social psychologists studying the relative performance of people in their chosen professions want iso-

FIGURE 8-1 Win Total and Points Total of 14 Teams in the Tiffany Victoria Memorial Fantasy Football League, 2011–2012: Note the variability—teams scoring between 1,250 and 1,300 points had between five and 10 wins while teams scoring between 1,050 and 1,150 had between three and eight wins. The boxed teams had unusually higher or lower number of wins than can be expected based on their points total, as indicated by the large vertical distance from the trend line.

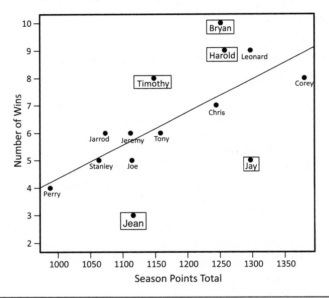

lated estimates of the effects of general intelligence, specific aptitudes, amount of experience, and personality traits. In modern theories of security returns, economists propose that prices fluctuate according to changes in factors such as economic growth and interest rates.

The knotty issue is untangling the factors. In any real-world situation, several factors in tandem bring about the observed outcome. But we want to examine the "all else being equal" scenarios, *ceteris paribus*, as economists like to call it. To rate coaching ability, the simplest thing to do is to take the points total: Corey was the best coach because his team scored the most points. However, this argument falls apart if fantasy performance reflects not just coaching ability but also other factors. If we hope to rate managerial acumen as well, we hit a snag as we cannot attribute the points total again. We therefore need to define two ratings that don't step over one another.

2. Living Out the Fantasy Life

Jay found an ominous note posted at ESPN.com the Friday approaching the second weekend of football in September 2012: "As long as the [Houston] Texans continue to view him as a game-time decision, then his status will need to be monitored closely up until kickoff, regardless of the precise nature of his injury." The note concerned Arian Foster, a productive running back who was Jay's first pick in the 2012 draft, the only time in many seasons he did not select a quarterback in this round (simply because other teams had taken all the desirable ones). Jay was in the same boat as a load of other FFL players who turned Foster into one of the most coveted fantasy RBs that season. It would pain him to have to deactivate the best player on the roster. Foster had been complaining about discomfort "around the knee area," and his team had yielded little about his status, hoping to catch their opponent unprepared.

Such game-time decisions are very upsetting to FFL players, Jay explained. If you're attending a live NFL game, it would take two brains and nimble fingers to follow the action on the field while scavenging on your smartphone for hearsay and half-truths from multiple websites about the upcoming games. If you're about to abandon your wife for most of Sunday, starting with the early game at noon and lasting until the end of *Sunday Night Football* close to midnight, now you're begging her forgiveness in the hours leading up to the first coin toss. Jay, despite living in the other hemisphere in Hong Kong, stayed up till the wee hours to confirm Foster's status. He activated Foster after learning he would start on Sunday. On other occasions, if he couldn't fight off sleepiness, he would be forced to forfeit possible points by selecting a "safe" player who would definitely play instead of the superstar who may sit out the game.

The work isn't done after the Sunday night game comes to a close, for the NFL schedules one or two Monday night matches. This is a key moment for any fantasy team who owns players eligible to play on Monday. Making the right moves on Monday may well turn a losing Sunday position golden, especially if you have more eligible players on your roster than your opponent.

After the game clock winds down on Monday night, another cycle begins. I asked Jay to describe his routine:

> I do a postmortem on my decisions the previous week. Did I pick the right guys on waivers? Did I start the right players? Did I follow the right advice from the right FFL news source? Then I assess my team: How did my players perform? What lessons can I learn from this week– matchups, player and coaching tendencies, etc.? Which players on waivers should I target? I know I put in more effort than most into this. The people with disadvantages are the ones who cannot adjust their lineups in the last hour . . . people who attend church, people who attend the football game, people in a far-off time zone, people who work the Sunday day shift. . . .

Week 2 reignited Jay's confidence in Hakeem Nicks, a wide receiver who had been one of his favorites in past seasons. Jay did not activate Nicks in the first two weeks because the receiver just recovered from foot surgery. Nicks scored three meager points in Week 1 against a strong defense but in Week 2, with the New York Giants having to overcome three first-half interceptions, they supercharged the passing game in the second half, leading to a monster day in which Nicks produced 25 fantasy points. Jay will not hesitate to insert Nicks in his lineups in future weeks.

One of the few players who might surpass even Jay in preparatory work is Leonard. Moreover, both Jay and Leonard play in two leagues simultaneously. During the 2012 draft, Leonard told the group he wasn't working presently, whereupon someone in the room quipped: "But fantasy football is your job!" Indeed.

3. A First Look at Coaching

Search for "Coach's Rating" on ESPN.com, and you will be served a weekly poll asking users to voice their approval or disapproval of each NFL head coach. In the FFL community, this style of rating, based on opinion rather than facts, makes few friends. Jay, Leonard, and many others spare no effort in their research, covering podcasts, TV shows, real-time chats, webcasts, Twitter feeds, Facebook messages, and so on. Also consulted are numerous websites catering to FFL players, such as ESPN, Yahoo!, Rotoworld.com, and FFtoday.com, which carry news, statistics, commentary, and projections. With so much data so easily accessed, why judge subjectively? Fantasy fans banter about the numbers in the peanut gallery, or at the negotiating table.

Entering the final week of the 2011 season, Perry and Jean co-owned the worst record in the league, 3 wins and 9 losses; both were aware that soon there would only be one team standing—eh, stranded—at the bottom. Jean tinkered with his lineup: For the two WRs, he had been rotating between Eric Decker, Julio Jones, and Early Doucet, and he chose Decker and Doucet; for defense, he selected the New England Patriots who would be facing the doormat Indianapolis Colts at home, instead of the New York Jets, a unit he favored in the first half of the season. As usual, Jean started two QBs, one of whom was Matt Hasselbeck, a 36-year-old veteran whose mediocre performance in 2011 unnerved Seattle Seahawks fans; the other QB was Carson Palmer.

By contrast, Perry activated the identical squad he had used in the previous three weeks; since he had lost three straight, this decision reflected either a white flag or a deep conviction. In the end, inaction brought victory. What doomed Jean was his unconditional trust in Hasselbeck. Needing only eight extra points to defeat Perry, if he had activated Jonathan Stewart as a second running back in place of the off-form Seattle QB, he would have won the last round with two points to spare; as it so happened Perry maxed out his points. Stewart is a competent back who scares fantasy coaches because he competes for playing time with several potent running threats on the Carolina Panthers and so his fantasy value is tied to the team's tactics, which vary from week to week. Jean gambled on Stewart during Week 11 and it paid off; he could have, and should have taken the same bet in Week 13.

The big idea is looking at what could have been in order to evaluate what was. Jean lost in Week 13 because he was outcoached. Perry played his best hand (74 points), but Jean could have scored 10 more points with just one swap—in fact, his maximum potential points were 86. (See Figure 8-2.)

In one sense, a good coach is able to pick out the nine players who would obtain the most fantasy points from the roster assembled by the manager. We can evaluate any selected squad by comparing it to the optimal squad. The points total relative to the attainable maximum for a given roster is what we call *Coach's Rating*. In Week 13, Jean was rated 78 percent, indicating his points total reached 78 percent of the potential; Perry scored a perfect 100, as he could not have done any better.

4. Another Look at Coaching

Tony, one of the founders of the Tiffany Victoria Memorial FFL, scored 71 points in Week 3 and 104 points in Week 4.

FIGURE 8-2   Jean's Selected Squad, a Modified Squad, and the Optimal Squad for Week 13 in the Tiffany Victoria Memorial Fantasy Football League, 2011–2012: Boxed selections could have improved the points total.

Position	Selected Squad	Modified Squad	Optimal Squad
QuarterBack	Carson Palmer	-- same --	-- same --
Running Back	Arian Foster	-- same --	-- same --
Wide Receiver 1	Eric Decker	-- same --	-- same --
Wide Receiver 2	Early Doucet	-- same --	Julio Jones
Tight End	Ed Dickson	-- same --	-- same --
Offense Wild Card	Matt Hasselbeck	Jonathan Stewart	Jonathan Stewart
Defense / Special Teams	Patriots D/ST	-- same --	Jets D/ST
Kicker	Jason Hanson	-- same --	-- same --
Head Coach	Packers Coach	-- same --	-- same --
Fantasy Points	67	77	86

His Coach's Ratings were in the 70s for both weeks, making them two of his least effective selections in 2011. This metric implies that the coach performed equally well in either week but in fact, Tony activated a truly wretched squad in Week 3. How do I know? Based on the 14 players he owned, I computed every one of the 256 squads Tony could have fielded in Week 3: The points totals fell into a tight range between 54 and 99, with the 71-point squad ahead of only 29 percent of the possibilities. Statistically speaking, 71 points was at the 29th percentile. For comparison, in Week 4, Tony's lineup ranked at the 66th percentile, between the worst squad at 59 points and the best at 133.

I call this rating the *Coach's "Prafs"* (*Percentile rank among feasible squads*). The Coach's Rating is a serviceable first approximation, and it's easier to obtain than the Coach's

Prafs, as it considers only the *optimal* squad. The Coach's Prafs looks at every possible squad, and is thus more telling, but it requires manipulating much more data.

Since I'll refer to the Coach's Prafs throughout the chapter, it helps to define the metric officially:

> Prafs is the percentile rank of the activated squad when compared to the range of points of all possible squads that can be constructed with the available roster. Its value is an integer between 0 and 100.

The coach who chooses the worst possible squad gets nada while the one who selects the optimal squad gets the maximum Coach's Prafs of 100. In the Tiffany Victoria Memorial FFL, the average weekly Coach's Prafs was 87 in 2011.

FIGURE 8-3 Coach's Prafs and Ranking in the Tiffany Victoria Memorial Fantasy Football League, 2011–2012: Cumulative Prafs takes values between 0 and 1,300. Five teams (inside the box) were bunched together when rated by coaching.

Total	Rank by Points Total			Coach's Prafs	Rank by Prafs
1380	1	Corey	Leonard	1214	1
1297	2	Leonard	Corey	1208	2
1297	3	Jay	Bryan	1200	3
1257	4	Harold	Chris	1182	4
1251	5	Bryan	Jarrod	1157	5
1244	6	Chris	Joe	1157	6
1158	7	Tony	Perry	1148	7
1148	8	Timothy	Stanley	1145	8
1116	9	Jean	Jay	1141	9
1114	10	Joe	Timothy	1120	10
1112	11	Jeremy	Jean	1086	11
1073	12	Jarrod	Tony	1064	12
1063	13	Stanley	Jeremy	1018	13
988	14	Perry	Harold	984	14

Thus, the league-average coach picked a squad that beat 87 percent of feasible squads. That was quite a competitive league! Jay and I trust this data-driven rating much more than ESPN's approval rating.

According to the cumulative Coach's Prafs, computed as the sum of weekly Coach's Prafs, the league's top coaches in 2011 were Leonard, Corey, Bryan, and Chris while Harold did rather worse than most. Jay ranked ninth in coaching skill, but a mere 16 points separated him from the fifth-ranked Jarrod, as the cumulative Coach's Prafs of five teams bunched up at around 1,150.

While defining the Coach's Prafs, I sneaked in a very important conditional: *The roster decisions are taken out of the coach's hands.* Like the restaurant critic Bruni, I zoomed in on how well the chef handled preset ingredients. It's as if playing hosts of the popular foodie show *Chopped*, in which contestants are challenged with concocting meals out of incongruous produce revealed only at game time. In a recent show, the kitchens gamely cooked up main courses using peanut butter, pork tenderloin, okra, and canned shrimp. Fixing the ingredients allows us to separate the effect of coaching/cooking from that of managing/shopping. We next turn our attention to managerial acumen.

5. Why Jay Ignored his Own Advice

An annual Fantasy Football League (FFL) ritual is the *draft* in which players build their season-opening rosters. This is when you grab the future Hall-of-Fame quarterbacks, stud running backs, or whoever you fancy when your turn arrives. In the Tiffany Victoria Memorial FFL, Jean, one of the cofounders, hosts this all-important event. (Although in 2012, the baton passed to Harold because Jean just closed on a new house.) Everyone except Jay and one other player attended in person. Jay called in via Skype from Hong Kong. The other player relayed his picks to Tony, the second cofounder, via the phone.

Someone screamed: "Malcom Floyd, wide receiver number 29 on the sheet!" As if playing Bingo, a bunch of players immediately crossed out the corresponding line on their cheat sheets. You can't pick Floyd when someone else already owns him. The *cheat sheet* lists all the available NFL players, their positions, and suggestions from the maker about who to select. Armed with pre-draft research, Jay is one of the few who brings his own cheat sheet, which he believes has better organization, and more recent data.

His first pick was Arian Foster, RB. No worthy quarterbacks were available by Jay's turn, as people made a bank run for QBs, listening to the advice of several fantasy football experts. In the second round, he gambled on Michael Vick, QB—no one else this year wanted to expend a high pick on Vick, who can be brilliant but is inconsistent and often injured. (Vick threw four interceptions but also tossed the game-winning touchdown in Week 1.)

Five hours and 10 rounds into the draft, Jay called it quits. The guys in Millbrae, California, were chattering about pizza orders and other random topics. The venue change delayed the start of the draft by almost two hours. Jay was up since 6 a.m., Hong Kong time, on a Sunday. The drifting of Skyped voices in and out of audible range, and the induced hunger from the jabber about food were too much to bear, so Jay appointed Tony to be his proxy and gave him instructions for the remaining rounds.

Jay's instruction for the last round was Greg Zuerlein, a rookie kicker on the St. Louis Rams. Normally, Jay shies away from rookies because of the risk. But being a Rams fan has its advantages: He heard that Zuerlein has a "huge leg," and he also expects the Rams to attempt many field goals as their offense struggles. Besides, the scoring formula in the Tiffany Victoria Memorial FFL rewards two extra points for long field goals of over 50 yards.

With all rosters set, the league is ready for the 2012 NFL season. The manager's job has barely begun. Leonard is one of the most serious and most successful players in the league. Ten years since joining, he has reached the final five times and won thrice. He reminds everyone that the count should have been four out of five since he was robbed in 2009. Fantasy players can get robbed by the referees, too. That year, Leonard was the league champion for one day, and then the NFL awarded his opponent Bryan an extra sack, which flipped the outcome.

Every Wednesday night, Leonard stalks the *waiver wire*, the list of players who haven't been drafted or have been dropped by other teams. It's like "waiting for Santa Claus to see if you've got what you wanted." Leonard manages his team intensely, making frequent adds and drops. The hiatus from work has energized his devotion to the pastime he adopted years ago to kick his gambling habit. Being able to react more quickly than others is his weapon. He watches every fantasy football show on TV, and monitors Android apps all day, all night. His team is named after his two favorite things in the world, the 49ers, and medical marijuana.

It was Tiffany Victoria who showed Leonard the ropes. Not only did Tiffany comanage the league, but she was also a formidable competitor. Leonard and Corey had the distinction of getting "beat by a girl" in a championship game. Tiffany commanded her bully pulpit, issuing entertaining dispatches each week giving "a girl's point of view" with illustrations. Leonard noticed how she weaved magic working the waiver wire. Sadly, her presence is reduced to memories.

6. Boxed in by Managers

Let's return to the big idea. What could have been is captured by the set of all feasible squads. These squads must satisfy the rules of the league, such as one or two quarterbacks (QBs) and two or three wide receivers (WRs). Take Perry's team in Week 8. He could have picked any one of 240 possible squads. Those squads would have scored as low as 18 points and as high as 67 points. The one he selected attained 62 points, just a hair below the maximum, and not surprisingly, Perry's Coach's Prafs was 98. (See Figure 8-4.)

Jay, Perry's opponent that week, also coached marvelously, earning Coach's Prafs of 99. He picked a squad that would have bested all but 1 percent of the 204 feasible squads. But the two well-coached teams found divergent fortunes, as Jay won this round handily by 90 to 62 points. Coaching alone could not explain this margin of difference: This is where managerial acumen looms. Notice that even Perry's best possible squad would have gotten only 67 points; meanwhile, the finest of Jay's would produce 92 points. Perry's managerial work placed his coach into a much tighter box than did Jay's.

I find Figure 8-5 helpful in understanding Bill Parcells's famous remark: "If they want you to cook the dinner, at least they ought to let you shop for some of the groceries." Each dot in the chart displays the points total for a feasible squad that could have been activated by Perry in a particular week

---

FIGURE 8-4 The Points Totals of All 240 Feasible Squads in Week 8 for Perry's Team in the Tiffany Victoria Memorial Fantasy Football League, 2011–2012: Each dot represents a squad, and a stack of dots represents multiple squads that would have earned the same points total. The large circle denotes the actual squad fielded by Perry, which would have beaten 98 percent of all feasible squads. Thus, Perry's Coach's Prafs was 98.

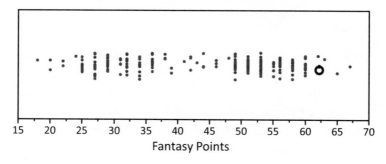

FIGURE 8-5 The Points Total of All Feasible Squads in All Weeks for Perry's Team in the Tiffany Victoria Memorial Fantasy Football League, 2011–2012: For each week, the range of dots traces the range of points that could have been earned by Perry. Managerial acumen is captured by the location and width of the range of dots. Coaching ability is evident in the location of large black dots (the selected squads) relative to the range of dots. For example, Perry coached well but managed poorly in Week 11; he neither managed nor coached well in Week 4.

of the season. Every possibility has been included. Head coaches feel like they are trapped in a box, and this box is visualized as the horizontal range of the dots in any given week. In Week 1, for instance, Perry's points total would fall between 62 and 113 points; no amount of maneuvering by the coach would change that fact. It is the general manager (GM), whose trades and deals determine the edges of this range. In some weeks, the GM ruined the team's chances; look at Week 11, when the maximum was a paltry 44 points and the minimum was 21. In other weeks, such as Week 7, the GM constructed a roster with far richer promise.

Frank Bruni's perspective is also visible here. No matter the ingredients, the chef's task is to elevate them. Focus for the moment on the large black dots in Figure 8-5, as they represent the actual squads activated by Perry each week. The further the large black dot leans toward the maximum point on the right side, the stronger was the coach's performance that week. Perry coached well in Weeks 6 and 8 relative to Weeks 4 and 7.

To get to a Manager's Rating, we need to equalize coaching in some manner. Here is one way to accomplish this. Concoct a "league-average coach," and ask what this imaginary coach would have attained given the week's roster. As mentioned before, the average coach in the Tiffany Victoria Memorial FFL played an 87th percentile squad. Comparing managers is thus the same as comparing 87th percentile squads. In Week 8, a league-average coach would have scored 58 points using Perry's roster, and 76 points using Jay's. Thus, Jay's managerial advantage amounted to 18 points. It was managing rather than coaching that gave Jay the win against Perry.

My metric for managerial acumen is called the Manager's "Polac" (Points obtained by league-average coach), and it refers to:

The points total that would have resulted if the team had employed a league-average coach, that is to say, one who plays the 87th percentile squad. This averaging effectively neutralizes the differential coaching abilities.

The managerial acumen of Jay and his fellow owners is sorted from best to worst according to cumulative Manager's Polac, which is just the sum of weekly Manager's Polac.

With our two metrics, we can now compare and contrast the 14 teams in the league. See Figure 8-6. We find three types of teams:

- The "All-Rounders," who both manage and coach well; they included Leonard, Corey, Bryan, Chris, and Jay.
- The "Motivators," who are above-average coaches but suffer from subpar roster decisions; Joe, Jarrod, Stanley,

---

FIGURE 8-6 Manager's Polac Points and Ranking in the Tiffany Victoria Memorial Fantasy Football League, 2011–2012: Polac is the points total that would have been obtained by the league-average coach given the available rosters.

Points Total	Rank by Points Total			Manager's Polac	Rank by Polac
1380	1	Corey	Harold	1275	1
1297	2	Leonard	Corey	1260	2
1297	3	Jay	Jay	1243	3
1257	4	Harold	Leonard	1187	4
1251	5	Bryan	Chris	1179	5
1244	6	Chris	Bryan	1150	6
1158	7	Tony	Jeremy	1137	7
1148	8	Timothy	Timothy	1127	8
1116	9	Jean	Tony	1121	9
1114	10	Joe	Jean	1115	10
1112	11	Jeremy	Joe	1037	11
1073	12	Jarrod	Jarrod	1013	12
1063	13	Stanley	Stanley	982	13
988	14	Perry	Perry	945	14

and Perry belonged here.

- The "Bean Counters," who create above-average rosters but are let down by coaching; Tony, Timothy, Jean, and Jeremy were in this category.

The odd man out or outlier in the league was Harold. He was the best manager and the worst coach in the league. (See Figure 8-7.)

7. Destiny

"The legend gets it right again," Jay noted, nodding as he took in the scatter plot. "Good coaching won't overcome poor

FIGURE 8-7 The 14 Teams in the Tiffany Victoria Memorial Fantasy Football League Divided into Three Types, According to Coaching and Managerial Skills: All-Rounders, who had above-average coaching and managerial skills; Bean Counters, who were let down by subpar coaching; and Motivators, who were let down by subpar managing. Harold was an outlier.

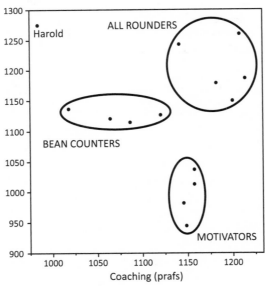

management. Good managers will beat good coaches in this league." He had in his head the overall performance of the 14 teams. The All-Rounders rising to the top was expected. What caught Jay's attention was the Bean Counters decisively outperforming the Motivators, who occupied four of the bottom five slots in terms of points total. Harold, the outlier, finished fourth in the league despite underwhelming coaching—on average, he activated the 76th percentile squad, markedly below the league average of 87th percentile.

A similar story unfolds if we use managerial acumen and coaching ability to explain win-loss records. In any week, if Jay attains any kind of managerial advantage, his chance of winning the matchup exceeds 80 percent. If his Manager's Polac is deficient by two or more points, he has an 86 percent chance of losing. These statements disregard coaching abilities. Coaching does have an effect, but it's secondary and weaker. Horrible coaching, defined as a disadvantage in Coach's Prafs of 22 points or more, negates any managerial advantage, curbing the chance of winning from 80 percent to 25 percent. Meanwhile, brilliant coaches who beat the competition by Coach's Prafs of 20 points or more work wonders with subpar rosters, turning an 86 percent chance of losing to a 62 percent chance of winning.

Something was not adding up, though. Jay was No. 3 by total points but third from last by total wins (tied with two other teams). And ultimately, wins count.

Could he have raised his standing by improving coaching? In a bit of a shock to us, more analysis showed he couldn't. Week 3 had Jay's worst Coach's Prafs of 71. That week, if he had swapped the Chargers defense (D) for the Cowboys defense (D), sat Shonn Greene at RB, and started an extra RB instead of an extra WR, he could have maxed out his points total at 120. But his opponent, Leonard, earned 141 points so Jay's fate was sealed. It was déjà vu in Week 2: Jay coached poorly with Coach's Prafs of 76 but his maximum potential of 113 points would still be one point below what Chris attained. Even at his best, Jay couldn't have beaten Corey in Week 6 either. In all three weeks, his opponents fielded their 99th percentile squads. You can imagine how Jay must have felt—like the shopper who bets on the wrong checkout line at the supermarket every time.

We started wondering if Tuff Toes were cursed in 2011 with a bad draw. The weekly matchups were randomly assigned at the start of the season. Each team gets to play every other team once. Teams have their ups and downs in terms of both managing the roster and fielding the game-day squad. This much is evident in the large weekly variance found in Figure 8-5. You'd like to face your opponents when they suffer a loss of form and avoid them when they hit a winning streak. You'd think luck might even out during the course of 13 weeks.

You'd think life is fair but for Jay, the inflexion point never came around in 2011. On average, his opposing managers achieved the Manager's Polac of 96 points when facing Tuff Toes. Together with Jean (who had the fewest wins in the league), Jay faced the toughest managers: Half of Jay's adversaries would have scored at least 98 points with the league-average coach. The opponents of Harold, by contrast, had an average managerial rating of 79 over the season, and only 20 percent of them scored 99 points or more. Harold's good fortune went a long way to explaining how he surmounted the worst-in-league coaching.

So far, we have focused on managing, since it makes a greater impact on the outcomes. Figure 8-8 looks at managing and coaching abilities simultaneously. Four of the top five teams with eight wins or more were dealt favorable matchups; many of their opponents showed up with subpar rosters. Bryan, who led the Tiffany Victoria Memorial FFL with 10 wins, enjoyed double fortune, as his opponents had below-average scores on both managing and coaching abilities. (A further analysis, using *regression*, demonstrates that win

FIGURE 8-8 Luck in the Tiffany Victoria Memorial Fantasy Football League, 2011–2012: Four of five teams with eight or more wins (circled below) faced opponents who managed poorly on average. Each number in the chart represents a team, and its label is the team's win total in 2011–2012. If every team experienced the same luck, the teams would cluster toward the middle of the chart.

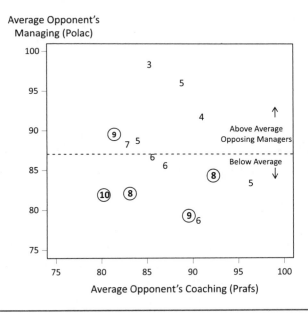

totals were more strongly correlated with their opponents' ratings, which measure luck, than with their own ratings. Nevertheless, the model with both sets of ratings is preferred to the model with only one.)

Just like other games, chance plays a prominent role in determining outcomes. Since we can't control luck, we should focus on maximizing the points total every week, and hope other things fall into place. Investing more time in building stronger rosters is worth it. Depending on your risk tolerance, you may want to aim for rosters with higher upsides but be aware that such lineups are typically also more unpredictable.

8. Following On at Home?

In the beginning, Jay asked me a couple of questions:

- Why didn't he get more wins given his favorable points total?
- Where should he seek improvement?

We pulled in some data, and cooked up a variety of analyses. This starts with describing the proposed two-factor model to explain the variation in total points. We can now confirm that the proposal has merit. The Coach's Prafs and the Manager's Polac metrics measure two distinct skills. If both metrics measured the same thing, which might be called "general fantasy football aptitude," they would have been almost perfectly correlated, and it wouldn't be possible to distinguish the Motivators (better at coaching) from the Bean Counters (better at managing). Though both factors are useful, managerial acumen has more impact than that of coaching.

The two-factor model is enhanced by adding the luck factor. Serendipity materializes in the average quality of the opponents one encounters during the season. With a random schedule, one expects each team to face similar levels of competition. But 13 weeks sometimes aren't enough for fairness to surface. Luck spices up the contest.

In many arenas today, such as fantasy sports, much data is publicly available. Frequently, the data help us answer tricky questions, and free us from speculation. Like the television chef, I showed off the pre-made stew standing in the oven. Now it's time to reveal the recipe.

Any number of websites host Fantasy Football Leagues. According to the formula for one's league, they furnish the basic ingredients: the weekly matchups, which teams won, and how many points each squad obtained.

Then, the key data set to assemble is the set of all feasible squads for each team each week. If we know the available rosters, we can compute the possible squads. We'd need statistics on how every NFL player performed in real life each week, which is easy to find online. For the Tiffany Victoria Memorial Fantasy Football League, each team has about 200 to 400 squads per week from which one is chosen. After every possibility is evaluated, we have generated quite a bit of data. These counterfactual scores are the key to the analysis. Jay and I believe in learning from past mistakes: Fantasy sports data is marvelously complete for this purpose. In real life, it is not possible to know what could have happened if, say, an NFL team had started one running back (RB) and not the other. It is trendy to carry over *Moneyball*-style analysis from real life to fantasyland. It is also smart to use counterfactual data that only FFL team owners could generate!

Chefs draw up the dinner menu while shopping for groceries. The smells and sights strike their fancy. Above all, they want to serve the freshest items in the market. They know that bad fish will ruin the best recipes. The chefs work out what goes well together with what. Back in the kitchen, they prepare the ingredients in a variety of ways: chopping, slicing, grinding, peeling, blanching, marinating, trimming, and so on. When cooking, they aim for a balance of taste, color, and aroma. Skewing one objective often spoils the dish, just like using mathematical models that omit parts of reality.

Analyzing data require similar skills:

- Having a clear head
- Knowing where and how to gather ingredients
- Having the flexibility to change courses
- Using creativity to shape the data
- Being vigilant about biases

Five years after his outburst, Bill Parcells eventually found a buyer of his vision. Leon Hess of the New York Jets told reporters: "I just want to be that little boy who goes along with him, and pushes the cart in the supermarket and let him fill it up."

Epilogue

Dear Reader,

I can't leave you with the idea that everyone must become data analysts to survive the era of Big Data. That is not the logical conclusion of this book. I do warn you that the wide availability of data will bring confusion and invite mischief. I hope you won't take data at face value ever again, and you see the power of looking under the hood.

- When a college president plays the steroids-didn't-help-us card, you'd recognize the attempt to trivialize the fraud.
- When medical researchers blame the failure to tame obesity on a bad metric, you'd whistle the stalling play.
- When Groupon's bankers play up free advertising to promote sales, you'd ask the counterfactual question.
- When a modeler claims his predictive model is so accurate it's creepy, you'd request the false-positive rate.
- If an expert denies her model uses any theoretical assumptions, you'd tune off.
- When an economist alleges bad weather brings bad economic data, you'd find out how things are counted.
- When a reporter cites raw, untampered-with economic data, you'd know not to compare individual months.

- When you develop a hypothesis about, for example, what factors affect fantasy-sports performance, you'd figure out what data you'd need and ask the right questions.

For the most part, you shouldn't have to handle the data. None of us have time to verify all claims, big and small. Knowing where the numbers come from will take you far. Understanding when and why assumptions are made is equally important. My goal is to offer you a tour of the back room—to show you how the numbers are made.

To that end, the book concludes with two episodes in the daily life of a data scientist. Google's Chief Economist Hal Varian bills this as a "sexy" job. What lies behind the "glamour"? I received many positive comments when the first piece appeared on my *Numbers Rule the World* blog. It is reproduced here, lightly edited, while the second piece is completely new.

1. Three Hours in the Life of a (Glorified) Data Scientist

A little puzzle gobbled three hours of my life one week, and not for the first time. The original task was to transfer a set of account numbers from one database to another. (If you must know, it was from an SQL Server box to Teradata, but the story doesn't change for any pair of vendors.) It's a fact of life that I'm always moving data from place to place. Those account numbers represent anonymous customers whose behavior I was interested in understanding. I'd extract all their interactions with my company up to a certain cutoff date. This is the kind of tasks that Target's modelers, from Chapter 5, did to predict pregnancy.

Within minutes, it became clear that the real task was getting Teradata to recognize a column of dates as dates. The column looked like this: 07/20/2010, 07/25/2010, 08/01/2010...

What was the problem? Of course, anyone could see those were dates. Well, Teradata disagreed—and until and unless—

Teradata was fully convinced I had offered it a column of dates, it would refuse to proceed with my main task, which was to compare the imported dates with the cutoff date.

Teradata thought 07/20/2010 was a string of text, not dates. I tried the simplest solution first: cast(my_column as date). It wouldn't budge, complaining that my_column contained "invalid dates." Flipping open the manual, I learned that a valid way to use the cast function is cast('2010-07-20' as date). So, I needed first to convert 07/20/2010 into '2010-07-20'.

I fumbled around a bit as I learned that Teradata does not support many classes of solutions I'm familiar with, such as *regular expressions*, the *MDY-type function* (which creates a date using month, day, and year as inputs), and the *find-and-substitute function*. So I quit trying to be cute, and reluctantly did it by brute force, via *substring functions* and *concatenate functions*. (A substring function extracts a section of text while a concatenate function merges two pieces of text.)

I gave Teradata a test case: cast('2010-07-20' as date) produced the Teradata date 07/20/2010, exactly as I wanted. Yes, that looked the same as my input column but human eyes deceive: If the database proclaimed it not to be a date, then it was not a date.

Satisfied with the test result, I now substituted '2010-07-20' with the brute-force substring-concatenate expression. Surprisingly, it failed, complaining again of invalid dates. I fished out some samples of these rejected dates. On inspection, they looked like dates. Smelled like dates.

Undeterred, I set aside the cast-as-date function, and applied the substring-concatenate expression to the column of dates; this ran without a hitch. As soon as I put the cast-as-date function back into the code, it bombed.

Now that the dumb but direct method stumbled, I went back to my cute ways. Maybe I could trick Teradata by splitting one step into two, first creating a new data table filled with the substring-concatenate output, the one operation that had worked so far, and then running the cast-as-date function on the new data.

Maybe not. No sooner had I placed the substring-concatenate expression together with two lines of code that generate a data table than it choked. The mystery deepened. The same code when used alone succeeded in producing Teradata dates, and yet as soon as I wrapped it inside table generation commands, it stalled. The error officially involved a missing something between the date variable and the comma sign inside the substring function. A so-called *syntax error*, as if I had violated a grammatical rule. This was irritating because the identical code ran smoothly when the output was sent to a pop-up window, but when the output was to be stored in a data table, the server apparently expected a different syntax! In any case, I couldn't figure out what Teradata was grumbling about.

Teradata and I were not friends at the moment. What to do? Like a spurned lover, I sought out my other good friend, the SQL Server box. What if I converted the column of dates to dates *before* transferring the data into Teradata?

I did that. After a laborious procedure, the data migrated to Teradata. Alas, the dates still showed up as strings of text. So, I doubled back to the SQL Server box: There, the dates were dates. This meant the program used to transfer data between the two platforms must have interpreted those as text.

My colleague suggested a Hail Mary. (Yes, now two "data scientists" were collaborating on this glamorous problem.) I was to force the dates to a "datetime" format, which looked like this: 07/20/2010 00:00:00. The time component was all zeroes, since the system never recorded information about time. Yes, I appended garbage to good data. It was a Hail Mary because we had no rationale why the database would read datetime but not date. You are reduced to doing crazier things when you run out of logical ideas.

It worked. It worked. It worked.

The SQL Server box turned the column of dates into

date times. Teradata not only read this correctly but read this three times, once as a datetime, once as a date, and once as a time.

I skipped over the exasperating data transfer procedure. Importing the data into Teradata required a special utility software. Half the time, the utility would not launch properly, and when that happened, I knew to issue an instruction to reset the software. On this particular day, the network connection was strained. After the utility opened, it took five attempts to find the network. One setting must be switched from its default value before running the utility. Modifying that setting always snipped the connection. So, it was another five tries to restore the connection. Only then could the data transfer get off the ground.

Three hours later, it worked. The "it" had morphed from finding customer activities to getting a database to recognize 07/20/2010 as a date, not text. The SQL Server box had closed up shop for the night, as it underwent maintenance. I still haven't found a single customer interaction. The project had a long way to go.

Every project entails situations like this. It's not an outlier. Welcome to the world of data science.

2. A Three-Day Face-off with 6,000 Words

Google has cemented itself as the gateway to the Internet. Raise your hand if you'd type "Fedex" into the search engine instead of entering fedex.com directly into the browser's address bar. Most websites derive the vast majority of visitors from Google searches. Google's algorithm is the king maker. When you enter a search term, the algorithm calculates which web pages are more relevant and shows you a list. The kingpin keeps marketers honest by tweaking its algorithm regularly. A recent major change turned on *Safe Search* for all U.S. users, a mode that suppresses searches for adult content using keywords like "xxx" or "boobs."

Many webmasters noticed an immediate drop in traffic from Google (they do not necessarily run adult websites, as the algorithm looks for relevant pages, not exact matches). On this Friday, my task was to estimate the impact of the Google tweak on my traffic. Including the weekend, I had at most three days to come to a conclusion. (Far from being a death sentence, you will see later that the time restriction was a lifesaver.)

A quick check showed incoming traffic from Google definitely declined. If my boss has no NUMBERSENSE, I can grant him the instant gratification he desires. Google modified the algorithm and the traffic plunged. Cause and effect. But he pays me to get better answers. Was the drop in visitors limited to adult search terms?

To figure this out, I had to investigate what people were searching for. I had to separately account for the adult traffic and the regular traffic. The tracking tool produced a list of 6,000 different search terms, ranked by their popularity, all of which sent visitors to my website in that month. A gaggle of questions rushed to my head while I stared at the data. I brushed aside the poor precision of only two significant digits (so, 2,500 visits appeared instead of 2,453). I convinced myself the tool returned all search terms, instead of a selection.

For the moment, one problem stumped me. The count of visits summed over all search terms did not match the traffic data from the other software. And it wasn't a 10 percent gap either: One number was *half* the size of the other. I have done this type of analysis enough to know the dirty secret of Web data—rather than this squeaky-clean, hyper-accountable system, it is the intractable web of thousands of entangled wires. Yay, Big Data. I have never seen a pair of tools that can compile comparable statistics; the word "identical" doesn't even exist in this realm. Still, it annoys me every time I see these gaps. It made me wonder, yet again, if the tracking tool presented only a sample of search terms. For not the only time today, I glanced at the clock, and gave myself 10 minutes

to investigate. Not a second more. Chatting with a few engi-
neers yielded few clues. It even felt stupid asking questions
since most people in the business have accepted the imprecision. Predictably, the minutes dripped away. I retreated to
theory. Here then: The *keyword tracker* produces accurate
estimates of relative shifts in traffic while the other software
is more reliable for counting aggregate traffic. These assumptions cannot be validated; that's why they are assumptions.
Sad but true, theory enters an analysis most often to paper
over water leaks.

I got back on track, now facing the most taxing piece of
work: dealing with the 6,000 search terms. I reckoned the
keywords had to drop into five large categories, or eight. My
will bent just imagining the hours ahead: read a word, assign
a label, read a word, assign a label, read, label, read, label. In
this moment of weakness, a shortcut seized me as a flu virus
invades your cells. Why not pull out the Top 100 words for
two separate months, and calculate the gain or loss in visits
for each word? This isn't anything groundbreaking—you've
seen this sort of analysis before.

It was also a trapdoor leading into a cul-de-sac. Like
many analytic plans, its flaws are well-hidden until you get
your hands dirty. Some 40 percent of the top search terms
featured in only one of the two months. Web search is a dy-
namic activity, and search terms come and go. Mitt Romney
was a top keyword in November 2012, but had faded by January 2013. Another problem: The Top 100 explained only 10
percent of total visits to the website. The analysis would miss
nine out of 10 visits. Popular searches are usually associated
with general keywords ("Halloween," "Harry Potter," etc.),
while the majority of visits come from Google users conducting more specific searches. This is known as the *long tail*: Lots
of little searches add up to a lot of visits.

All of that used up an hour, but I felt a positive vibe. If
you're following, you'd notice I really had no end product, as
I was about to abandon the shortcut. The detour made the

original plan more palatable. There was no getting around labeling the 6,000 keywords. The earlier problems would then evaporate. Categories persist from month to month, even if the search terms vary. Also, all the visits would be accounted for, not just 10 percent. I could visualize the output now, a table showing the trend in visits for each category of search terms.

I lied if I said I had screened all 6,000 search terms. Right from the start, that was an impossibility. If I were a robot that could process one keyword every 10 seconds, it would have taken more than 16 hours for the entire list, assuming no rest at all, and no lapse in concentration. This was where the time constraint brought a measure of sanity. I classified as many words as I could within the allotted time. The actual experience of labeling each search term was strangely hypnotic. *"The Walking Dead"* is a television show. "Xtube" is an adult website. "Manchester United v Chelsea" is a sports event. It was mindless, repetitive work, but it put my brain at ease. I had to force myself to stop for dinner.

The analytic plan gained my trust as the exercise rolled on. Carrying on the Top 100 analysis would have been foolish. The tracking tool—developed by a large, respected company—delivers the search terms *raw*, the exact words typed in by Google users, typing errors and the works. There were at least 20 misspellings of my company's name. Any popular search existed in numerous permutations: "Chelsea v Manchester United," "Chelsea v ManU," "download Man U vs Chelsea," and so forth. Without larger categories, the information would be lost in small slices.

Over the next day, I got into a routine. I was a laborer. In front of me lay two big buckets, the bucket of the classified and the bucket of the vast unclassified. I kept pulling stuff out of the second bucket, and shoveling them over to the first one. Back and forth I went. Every hour or so, I broke out of hypnosis and wondered if I'd done enough. You see, the long tail had me by the neck. The first 100 terms were the easiest.

The next 100 got harder progressively, as each one explained less of my traffic. Also, the keywords became less familiar, so I often had to look up the right label. Not only was the shovel shrinking all the time, my movement also slowed. If I stopped before having classified enough terms, the output would suffer the same shortcomings as the Top 100 analysis.

Are you worried for my mental state? Why doesn't he program a computer to do this grunt work? That thought crossed my mind, too. It's quite a shock to realize that all the progress we've made in information technologies has not resulted in an automated solution. In fact, today's computers do not understand language. All they do is match text: They can tell me whether the words "empirical Bayes model" are found on a specific Web page. Computers can't figure out that this Web page is about statistical methods, unless they are specifically trained for this classification task, which means learning from examples of Web pages with correct labels. I could spend time building this training data set, or I could just go further, and finish the analysis.

It's unlikely the computer could solve some of the trickier issues I came across. Are there one or two categories amongst the search terms "the dutch," "the dutch nyc," and "the dutch brunch"? The last two keywords belong together, referring to a hot new restaurant in Manhattan. The tracking tool indicated 1,200 searches for "the dutch." Were these people looking for the restaurant or the people of the Netherlands? It's probably some of both, and no computer or human can disentangle that without supplementary data.

I finally called a time-out on Sunday. The remaining, unclassified keywords appeared immaterial, each contributing only hundreds of visits. To my surprise, the count of labeled words reached only 1,000, and only accounted for half of the website's traffic. All those hours of hard labor, and so much unfinished business! Thankfully, the analysis confirmed that the stopping time was chosen wisely. While I only managed to label half of the traffic, the unclassified half, now lumped into the uninformative "Others" category, contributed a miniscule proportion of online sales.

The exercise began with 12,000 numbers, the traffic generated by 6,000 search terms during two months. A spreadsheet containing all of that data, or even one reduced to just the Top 100 keywords, confuses the audience. Three days later, I had everything summarized on one page, the 6,000 keywords dropped into six categories, and for each, I knew the rate of decline. Was the drop in visitors limited to adult search terms? I discovered that the Google tweak suppressed a fair amount of searches for explicit content, but also wiped out some other categories.

References

Prologue

Fung, Kaiser, *Numbers Rule Your World: The Hidden Influence of Probability and Statistics on Everything You Do*, New York: McGraw-Hill, 2010.

Gelman, Andrew, "Causality and Statistical Learning," *Statistical Modeling, Causal Inference, and Social Science* blog, March 27, 2013. Contains further examples of data interpretation.

Ioannidis, John, P.A., "Genetic Associations: True or False?" *Trends in Molecular Medicine* 9, no. 4(April 2009): 135–138. One of a series of articles about false-positive results in peer-reviewed publications.

Kaushik, Avinash, *Web Analytics 2.0: The Art of Online Accountability and Science of Customer Centricity*, New York: Wiley, 2010. Useful introduction to Web data.

McKinsey Global Institute, "Big Data: The Next Frontier for Innovation, Competition, and Productivity," June 2011.

Pollster.com blog. "Rasmussen, Massachusetts, and Party ID," blog entry by Mark Blumenthal, Jan. 6, 2010.

RealClearPolitics.com has national poll data.

Shaw, Linda, wrote several articles in *Seattle Times* about small schools in Washington state.

Silver, Nate, *The Signal and the Noise: Why Most Predictions Fail and Some Don't*, New York: Penguin, 2012.

UnskewedPolls.com has national poll data.

Wainer, Howard, and Harris L. Zwerling, "Evidence That Small Schools Do Not Improve Student Achievement," *Phi Delta Kappan* 88, no. 4(Dec. 2006): 300–303.

1 • Law School Ranking

Caples, John, and Fred Hahn, *Tested Advertising Methods*, 5th edition, New York: Prentice Hall, 1988. Classic book on direct marketing.

Caron, Paul L., "Did 16 Law Schools Commit Rankings Malpractice?" *Tax Prof* blog, May 12, 2010.

Gordon, Larry, "Claremont McKenna College Inflated Freshman SAT Scores, Probe Finds," *Los Angeles Times*, Jan. 30, 2012.

Henderson, Bill, "If Yale is #1 in *U.S. News*, Is It the Best Law School?" *Empirical Legal Studies* blog, Sept. 2, 2008.

Jones Day, and Duff & Phelps, "Investigative Report: University of Illinois College of Law Class Profile Reporting," Nov. 7, 2011.

Leiter, Brian, "Sextonism Watch," *Brian Leiter's Law School Reports* blog, Aug. 5, 2005.

Meisel, Hannah, "College of Law Report Reveals Pless's Interest in iLEAP Program," *Daily Illini*, Nov. 10, 2011.

Michigan School of Law Class Statistics can be found at http://www.law.umich.edu/prospectivestudents/Pages/classstatistics.aspx.

O'Melveny & Myers, LLP, Apalla Chopra, and Carolyn Kubota, "Investigative Report Prepared on Behalf of the Board of Trustees of Claremont McKenna College," April 17, 2012.

Sauder, Michael, and Wendy Nelson Espeland, "Strength in Numbers? The Advantages of Multiple Rankings," *Indiana Law Journal* 81, no. 205 (2006): 205–206.

Zearfoss, Sarah, "Revisiting the Wolverine Scholars Program," The University of Michigan Career Center blog, June 1, 2011.

The following three are good references on the flaws of the *U.S. News* methodology. But note that there does not exist some provably optimal ranking:

Cloud, Morgan, and George B. Shepherd, "Law Deans in Jail," *Emory University School of Law Legal Studies Research Paper Series* #12-199, 2012.

Klein, Stephen P., and Laura Hamilton, "The Validity of the *U.S. News and World Report* Ranking of ABA Law Schools," online at Association of American Law Schools website, www.aals.org, Feb. 18, 1998.

Seto, Theodore P., "Understanding the *U.S. News* Law School Rankings," *SMU Law Review* 60(2007): 493–576.

2 • Obesity

Centers for Disease Control (CDC) website. State-level obesity statistics and maps are available from http://www.cdc.gov/obesity/data/adult.html.

Curtin, Francois, Alfredo Morabia, Claude Pichard, and Daniel O. Slosman, "Body Mass Index Compared to Dual-Energy X-Ray Absorptiometry: Evidence for a Spectrum Bias," *Journal of Clinical Epidemiology* 50, no. 7(1997): 837–843.

Flegal, Katherine M., Margaret D. Carroll, Cynthia L. Ogden, Lester R. Curtin, "Prevalence and Trends in Obesity Among U.S. Adults, 1999–2008," *Journal of the American Medical Association* 303, no. 3(Jan. 20, 2010): 235–241. Flegal and associates have published a series of papers tracking obesity trends in the United States. I used their statistics to estimate the prevalence of obesity by gender and age group.

Fung, Kaiser, "The Inevitable Perversion of Measurement," *Numbers Rule Your World* blog, June 4, 2012.

Gallagher, Dympna, Steven B. Heymsfield, Moonseong Heo, Susan A. Jebb, Peter R. Murgatroyd, and Yoichi Sakamoto, "Healthy Percentage Body Fat Ranges: An Approach for Developing Guidelines Based on Body Mass Index," *American Journal of Clinical Nutrition* 72(2000): 694–670.

Healy, Melissa, "We May Be Fatter Than We Think, Researchers Report," *Los Angeles Times,* April 2, 2012, accessed online.

Keys, Ancel, Flaminio Fidanza, Martti J. Karvonen, Noboru Kimura, and Henry L. Taylor, "Indices of Relative Weight and Obesity," *Journal of Chronic Diseases* 25, no. 6(1972): 329–343. This is the paper that coined the Body Mass Index.

Klein, Samuel, David B. Allison, Steven B. Heymsfield, David E. Kelley, Rudolph L. Leibel, Cathy Nonas, and Richard Kahn, "Waist Circumference and Cardiometabolic Risk," *Diabetes Care* 30, no. 6(2007): 1647–1652.

Kopelman, Peter, "Foresight Report: The Obesity Challenge Ahead," *Proceedings of the Nutrition Society* 69(2009): 80–85. British research into causes of obesity.

Kurth, Tobias, J. Michael Gaziano, Klaus Berger, Carlos S. Kase, Kathryn M. Rexrode, Nancy R. Cook, Julie E. Buring, and JoAnn E. Manson, "Body Mass Index and the Risk of Stroke in Men," *Archives of Internal Medicine* 162, no. 22(2002): 2557–2562. Relative risk is the probability of disease of the exposed population relative to that of a baseline population.

Ogden, Cynthia L., Susan Z. Yanovski, Margaret D. Carroll, and Katherine M. Flegal, "The Epidemiology of Obesity," *Gastroenterology* 132(2007): 2087–2102.

Prentice, A. M., and S. A. Jebb, "Beyond Body Mass Index," *Obesity Reviews* 2(2001): 141–147. Good summary of the case against BMI.

Shah, Nirav R., and Eric R. Braverman, "Measuring Adiposity in Patients: The Utility of Body Mass Index (BMI), Percent Body Fat, and Leptin," *PLoS ONE* 7, no. 4(April 2012): e33308.

Taubes, Gary, *Why We Get Fat and What to Do About It*, New York: Knopf, 2011. A science reporter criticizes the medical establishment's approach to the obesity epidemic.

The Weight of the Nation, HBO, May 2012.

Winfrey, Oprah, "How Did I Let This Happen Again?" *O, The Oprah Magazine*, Jan. 2009.

3 and 4 • Groupon

All Things Digital blog. The Andrew Mason memo was leaked to Kara Swisher.

Arrington, Michael, "LivingSocial Financials Exposed: $2.9 Billion Valuation, $50 Million in Revenue Per Month," *Tech Crunch* blog, Apr. 15, 2010.

Burker, Jessie, "Groupon in Retrospect," *Posies Cafe* blog, Sept. 11, 2010.

Eaton, Kit, "Twitter Really Works: Makes $6.5 Million in Sales for Dell," *Fast Company* blog, Dec. 8, 2009.

Salmon, Felix, "Grouponomics," *Felix Salmon Reuters* blog, May 4, 2011.

———; "Whither Groupon?" *Felix Salmon Reuters* blog, Sept. 1, 2011.

Fung, Kaiser, "Grouponomics, and the Power of Counterfactual Thinking," *Numbers Rule Your World* blog, May 6, 2011.

———, *Numbers Rule Your World: The Hidden Influence of Probability and Statistics on Everything You Do*, New York: McGraw-Hill, 2010, Chapter 4.

Groupon, Inc., Initial Public Offering Prospectus (Form S-1), June 2011.

———, Fourth Quarter 2012 Results, Feb. 27, 2013.

IDC and Business Software Alliance, "Seventh Annual BSA/IDC Global Software 09 Piracy Study," May 2010.

Pogue, David, "Psyched to Buy, in Groups," *New York Times*, p. B1, Feb. 10, 2011.

Popper, Ben, "Greed is Groupon: Can Anyone Save the Company From Itself?" *The Verge*, March 13, 2013. One of many postmortem accounts of the Groupon story.

Primack, Dan, "What Really Happened at LivingSocial?" *Fortune The Term Sheet* blog, Feb. 21, 2013.

Rubin, Donald B., "Causal Inference Using Potential Outcomes," *Journal of the American Statistical Association* 100, no. 469(2005): 322–331.

Siegel, Eric, *Predictive Analytics: The Power to Predict Who Will Click, Buy, Lie, or Die*, New York: Wiley, 2013. Read particularly the chapter on net lift models.

5 • Target

Anderson, Chris, "The End of Theory: How the Data Deluge Makes the Scientific Method Obsolete," *Wired*, June 23, 2008.

Zhong, Chen-Bo, and Katie Liljenquist, "Washing Away Your Sins: Threatened Morality and Physical Cleansing," *Science* 313, no. 5792(Sept. 8, 2006):1451–1452.

Derman, Emanuel, *Models.Behaving.Badly: Why Confusing Illusion with Reality Can Lead to Disaster, on Wall Street and in Life*, New York: Free Press, 2012. Good discussion of how models of social science differ from models of physics.

Duhigg, Charles, *The Power of Habit: Why We Do What We Do in Life and Business*, New York: Random House, 2012.

——; "How Companies Learn Your Secrets," *New York Times Magazine*, Feb. 16, 2012.

Fung, Kaiser, *Numbers Rule Your World: The Hidden Influence of Probability and Statistics on Everything You Do*, New York: McGraw-Hill, 2010, Chapter 4.

Kahneman, Daniel, *Thinking, Fast and Slow*, 2011, New York: Farrar, Straus, and Giroux, 2011.

Taleb, Nassim Nicholas, *The Black Swan: The Impact of the Highly Improbable*, New York: Random House, 2007. Discusses the hazard of overreliance on past patterns.

6 • Unemployment

Bowler, Mary, and Teresa L. Morisi, "Understanding the Employment Measures from the CPS and CES Survey," *Monthly Labor Review* 129: (Feb. 2006) 23–38.

Dahlin, Brian, "How the Business Birth/Death Model Improves Payroll Employment Estimates," *Issues in Labor Statistics*, Oct. 3, 2008.

——; "Seasonal Adjustment and Calendar Effects Treatment in All Employee Hours and Earnings Estimates," *Issues in Labor Statistics*, Feb. 5, 2010.

Haugen, Steven E., "Measures of Labor Underutilization from the Current Population Survey," *Bureau of Labor Statistics Working Paper* 424, March 2009.

Krugman, Paul, "Constant-Demography Employment," *The Conscience of a Liberal* blog, Oct. 6, 2012. An economist takes on how to adjust for population growth in measuring employment.

Mueller, Kirk, "Impact of Business Births and Deaths in the Payroll Survey," *Monthly Labor Review* 129: (May 2006) 28–34.

New York Post website. John Crudele's columns are online at www.nypost.com.

7 • Inflation

Bureau of Labor Statistics, *BLS Handbook of Methods*, Chapters 1, 16, and 17.

———, "The So-Called 'Core' Index: History and Uses of the Index for All Items Less Food and Energy," *Focus on Prices and Spending* 1(15): 1–3, Feb. 2011.

Bureau of Labor Statistics website. Data on consumer price indices and consumer expenditures are available at http://www.bls.gov/cpi and http://www.bls.gov/cex, respectively.

Dickson, Peter R., and Alan G. Sawyer, "The Price Knowledge and Search of Supermarket Shoppers," *Journal of Marketing* 54: 42–53, July 1990.

Kahneman, op. cit.

Merriam-Webster's Collegiate Dictionary, Eleventh Edition, Springfield, MA: Merriam-Webster, 2004, p. 277. Definition of the word "core."

Schnepf, Randy, "Consumers and Food Price Inflation," *Congressional Research Service Report* 7-5700, Oct. 4, 2012.

Van der Klaauw, Wilbert, Wandi Bruine de Bruin, Giorgio Topa, Simon Potter, and Michael Bryan, "Rethinking the Measurement of Household Inflation Expectations: Preliminary Findings," *Federal Reserve Bank of New York Staff Report* No. 359, Dec. 2008, Figure 1.2.

8 • Fantasy Football

ESPN.com. Fantasy football data can be obtained from websites that host leagues.

Fung, Kaiser, *Numbers Rule Your World: The Hidden Influence of Probability and Statistics on Everything You Do*, New York: McGraw-Hill, 2010, Chapter 3. The names of the league and team owners were altered.

Index